W9-BJC-679

Fiber Optics

Technology and Applications

Applications of Communications Theory
Series Editor: R. W. Lucky, *AT&T Bell Laboratories*

INTRODUCTION TO COMMUNICATION SCIENCE AND SYSTEMS
John R. Pierce and Edward C. Posner

OPTICAL FIBER TRANSMISSION SYSTEMS
Stewart D. Personick

TELECOMMUNICATIONS SWITCHING
J. Gordon Pearce

ERROR-CORRECTION CODING FOR DIGITAL COMMUNICATIONS
George C. Clark, Jr., and J. Bibb Cain

COMPUTER NETWORK ARCHITECTURES AND PROTOCOLS
Edited by Paul E. Green, Jr.

FUNDAMENTALS OF DIGITAL SWITCHING
Edited by John C. McDonald

DEEP SPACE TELECOMMUNICATIONS SYSTEMS ENGINEERING
Edited by Joseph H. Yuen

MODELING AND ANALYSIS OF COMPUTER
COMMUNICATIONS NETWORKS
Jeremiah F. Hayes

MODERN TELECOMMUNICATION
E. Bryan Carne

FIBER OPTICS: Technology and Applications
Stewart D. Personick

A Continuation Order Plan is available for this series. A continuation order will bring delivery of each new volume immediately upon publication. Volumes are billed only upon actual shipment. For further information please contact the publisher.

Fiber Optics

Technology and Applications

Stewart D. Personick

Bell Communications Research
Red Bank, New Jersey

PLENUM PRESS • NEW YORK AND LONDON

Library of Congress Cataloging in Publication Data

Personick, Stewart D.
Fiber optics.

(Applications of communications theory)
Includes bibliographies and index.
1. Fiber optics. I. Title. II. Series.
TA1800.P47 1985 621.36′92 85-12370
ISBN 0-306-42079-1

First Printing—August 1985
Second Printing—June 1986

A Division of Plenum Publishing Corporation
233 Spring Street, New York, N.Y. 10013

Printed in the United States of America

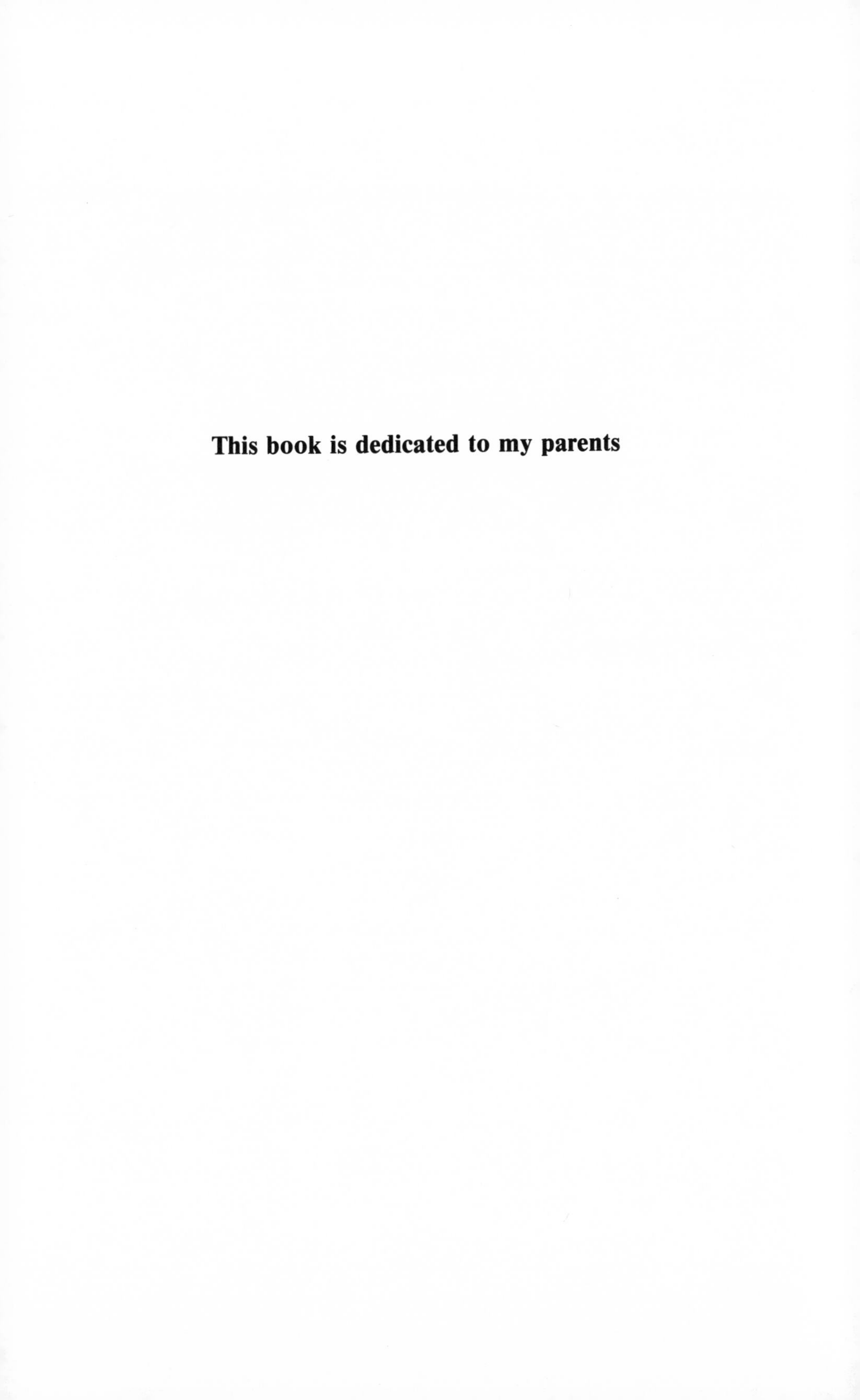

This book is dedicated to my parents

Preface

This book is an outgrowth of a course given by the author for people in industry, government, and universities wishing to understand the implications of emerging optical fiber technology, and how this technology can be applied to their specific information transport and sensing system needs. The course, in turn, is an outgrowth of 15 exciting years during which the author participated in the research and development, as well as in the application, of fiber technology. The aim of this book is to provide the reader with a working knowledge of the components and subsystems which make up fiber systems and of a wide variety of implemented and proposed applications for fiber technology. The book is directed primarily at those who would be users, as opposed to developers, of the technology.

The first half of this book is an overview of components and subsystems including fibers, connectors, cables, sources, detectors, receivers, transmitters, and miscellaneous components. The goal is to familiarize the reader with the properties of these components and subsystems to the extent necessary to understand their potential applications and limitations.

The second half of the book surveys a wide range of existing and hypothetical applications, illustrating what is being done with currently available components and subsystems, and what might be done as a result of future developments. Applications addressed include point-to-point digital transmission systems for telecommunications trunking and for data links between computing entities, analog point-to-point links for r.f., i.f., video, and broadband telemetry transport, local area networks for data and broadband networks for telecommunications integrated-services distribution, sensing systems, photonic switching, and several miscellaneous applications.

I would like to gratefully acknowledge Ms. Peggy Bergstrom, who did the text processing of this manuscript.

Contents

Part I: Technology

Part II: Applications

PART I: TECHNOLOGY

1

Technology Overview

The purpose of the sections which follow this technology overview are to familiarize the reader with the components and subsystems which make up fiber systems, to the extent necessary to understand the capabilities and limitations of these system building blocks. Before doing that, it is helpful to present a brief overview of fiber optics technology, in order that the somewhat more detailed discussions below can be put into context.

Figure 1-1 shows an illustration of a specific point-to-point telecommunication system which is implemented with fiber optic components. Beginning at the upper left we have a source of analog information, which is in this case a telephone user. The analog signal is converted to a sequence of binary digits with a conventional electronic terminal called a coder. The use of the coder for analog-to-digital conversion is not restricted in any way to fiber optic systems. The binary data stream emitted by the coder can be converted back into an analog signal by a complementary terminal called a decoder. If the coder and decoder are placed back-to-back, the sending and receiving users will experience some quality of communication which depends on the details of the coder and decoder designs. If the two users are not colocated, then one requires a digital transmission system to transport the binary data from the coder output to the decoder input. To the extent that the binary digits are transported without error, the performance of the overall communications link will be essentially the same as for back-to-back terminals. To implement the transmission function with optical fibers one needs, at a minimum, an optical transmitter, a fiber, and an optical receiver.

The optical transmitter converts electrical pulses into pulses of optical power. The fiber captures some fraction of the power emitted by the transmitter and transports it to the receiving end of the link. The optical receiver converts the pulses of optical power emerging from the fiber into

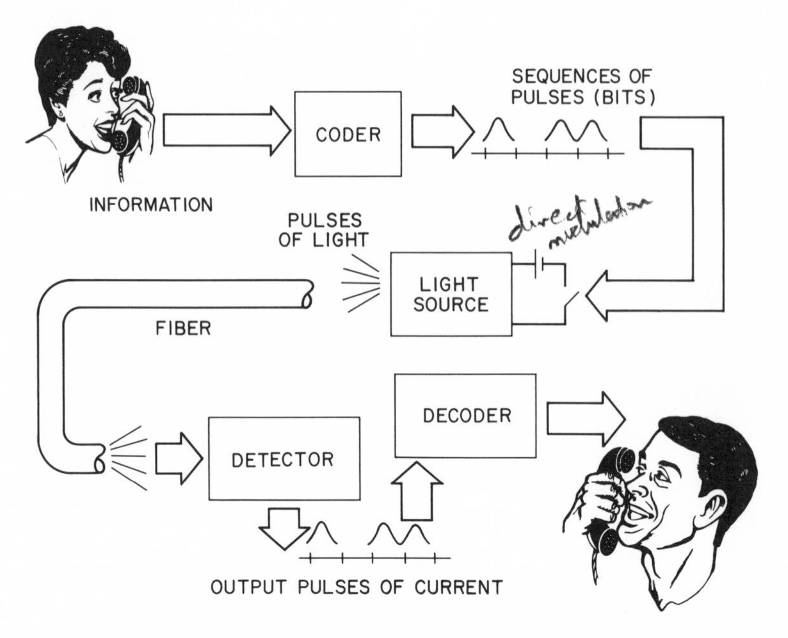

Figure 1-1 Example of a Point-to-Point Telecommunications System Implemented with Fiber Optic Technology.

electrical pulses. It is useful to note that although the electromagnetic field emitted by an optical source varies at optical rates (around 3×10^{14} Hz for typical fiber-optic-system emitters) power is a baseband quantity (always positive) which varies with the modulation. As shown in Figure 1-1, the optical power can be modulated by modulating the electrical current that supplies the light emitter. To the extent that the light output can follow variations in the electrical drive current at the required rates this "direct modulation" approach is the simplest to implement. Indeed, a very large percentage of systems using fiber technology apply this method. As we shall see in the in-depth discussions below, there are occasions where it is desirable or necessary to modulate the light with an external device after it is generated by the emitter.

Light which is emitted is captured, in part, and guided by the fiber by the process of total internal reflection. As this captured light propagates along the fiber it experiences "attenuation" and "delay distortion." Attenuation corresponds to a decay in the power level due to the conversion of propagating power into heat and due to scattering of propagating power from the fiber. Delay distortion corresponds to the smearing out in time of the modulation of the optical signal as it propagates. These effects will be described in more detail below. After some distance of propagation the effects of attenuation and delay distortion will accumulate to the point where the ability to extract the digital information from the modulated optical signal is jeopardized. At that point, the optical signal must be converted back to electrical form for processing to remove these accumulated degradations. An electrical regenerator produces a new, "cleaned-up," digital electrical bit stream which can drive a new optical transmitter or the decoder, as appropriate.

The implementation of fiber optic systems requires reliable, economical, optical sources (emitters), at appropriate wavelengths with appropriate input–output characteristics; cabled fibers with low attenuation and acceptable delay distortion characteristics; means for splicing and for implementing rearrangeable connections; appropriate high sensitivity optical detectors; as well as a good deal of electronic circuitry to interface to the optical components. Some systems require other components such as optical modulators, switches, couplers, multiplexers, etc. In the sections below we shall discuss these components in some detail.

- Shutters could be used instead of direct modulation (DM)
- Currently DM meets todays requirements.

2

Optical Fibers

The purpose of this chapter is to describe optical fibers in terms of their input–output characteristics; to describe the methods by which fibers are fabricated; and to review cabling, splicing, and connectorization techniques, which are key to the practical utilization of the fibers.

2.1 Input – Output Properties

Fibers transport light signals from place to place just as metallic conductors transport electrical signals. Fibers can guide light around bends and can carry light for long distances with remarkably little attenuation (loss of light power) [1–3]. Although the transmission characteristics of modern fibers defy previous intuition, they are not perfect. Fibers introduce light loss and smearing of the modulation imposed on the light signals to represent information. These attenuation and delay distortion effects (respectively) limit the distances which can be spanned without electro-optic repeaters (effectively amplifiers) and limit the information rates which can be carried over long distances as well.

2.1.1 Propagation Models [4, 5]

A thorough and relatively precise understanding of the way in which light signals propagate in optical fibers requires one to treat the fiber as a dielectric waveguide, and to use Maxwell's equations to calculate the fields which can exist within the fiber. Such analyses have been carried out and published in numerous references. Although we will draw upon a few results of these analyses a little later in this discussion, our purposes are served better by using an approximate (but adequate) model known in the field as geometrical optics.

The geometrical optics model of fiber propagation is illustrated in Figure 2-1. We begin with a simple model of a light source which emits "rays" of light from different points on its surface; and which travel at various angles relative to the reference axis shown. Consider ray 1. This ray impinges on the fiber core and enters the core region as shown. The core is made of a material (silica glass) whose optical density is higher than that of air. As a result of this, the ray entering the core is "refracted" or bent toward the reference axis. This can be explained by imagining a wave front perpendicular to the ray as shown in Figure 2-2. As the ray enters the higher-density medium (traveling right to left) the top portion of the wave front enters first. Since the velocity of light is lower in the denser medium, the top portion of the wave front also slows down first. This slowing action causes the bottom portion of the wave front to move ahead relative to the top portion. As a result the wave front and the direction of propagation turn toward the reference axis. Refer again to Figure 2-1. After traveling some distance in the core region, ray 1 strikes the core–cladding interface. The cladding has a slightly lower optical density than the core, and this causes the ray to be refracted toward the reference axis due to the same mechanism as was described above. Ray 1 then propagates in the cladding region until it strikes the interface between the cladding and the jacket. In this simplified model we shall assume that the ray is lost there, due to scattering at the relatively rough jacket–cladding interface, light absorption in the relatively lossy outermost portion of the cladding region, and field leakage into the highly absorptive jacket material. In actuality it is possible for the ray to be reflected at the cladding–jacket interface and to propagate a further distance. However, the ray will rapidly be lost as it propagates, due to the accumulation of these loss mechanisms; so our simplified model is adequate for this discussion.

Consider ray 2 in Figure 2-1. This ray strikes the core–cladding interface at a relatively shallow angle (relative to the reference axis). It can be shown that if the angle of incidence is sufficiently small (how small depends on the optical density difference between the core and the cladding, and will be quantified below) then the ray will not enter the cladding, but will be totally reflected instead. Thus the ray will propagate down the fiber following a zig-zag path as shown, reflecting at each incidence upon the core–cladding interface. Other rays are shown which are either incident upon the cladding region or which miss the fiber entirely. These rays are not captured and guided. Thus we see that some fraction of the light emitted by the simplified model source can be captured by the fiber. We can quantify this by using some results obtained from more detailed physical optics analyses.

Consider again Figures 2-1 and 2-2. The optical density of a material can be quantified in terms of its refractive index, n. We shall call the index of refraction of the core material n_{core}, and we shall call the index of refraction of the cladding material n_{cladding}. (Later in this section

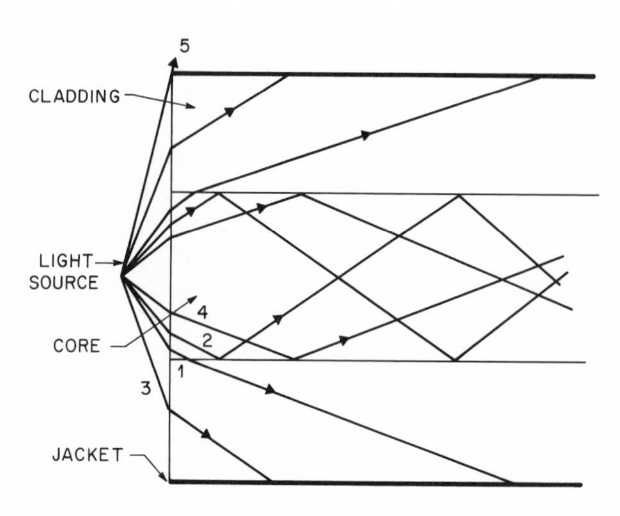

Figure 2-1 Geometrical Optics Model of Light Propagation in a Fiber.

we shall consider fibers where the index of refraction is not uniform in the core but varies at different radii from the reference axis. For the moment let us assume that the core index is uniform for all radii up to the core–cladding interface.) To understand quantitatively the amount of bending which a ray will experience in passing from one medium into another with a different index of refraction one can use various analytical methods. Some of these methods are more intuitive than others. Some

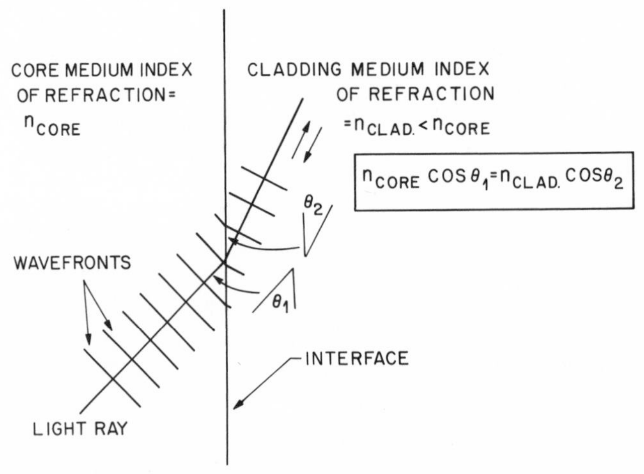

Figure 2-2 Illustration of Refraction at an Interface.

are based on assumptions which, although intuitive, can only be derived from other methods. We can start with Maxwell's equations, and postulate three propagating electromagnetic waves: one propagating in the core region toward the interface; one reflected from the interface and propagating in the core region; and one propagating away from the interface in the cladding region. We can then use boundary conditions to solve for the amplitudes and directions of such a set of three propagating waves, which together satisfy Maxwell's equations on both sides of the interface and at the interface. From such calculations we obtain Snell's law, which is often stated as a "given" in first year college physics courses. If we define the angle of the wave traveling in the core toward the interface (relative to the interface as shown in Figure 2-2) as θ_1, then according to Snell's law we have

$$n_{core} \cos(\theta_1) = n_{cladding} \cos(\theta_2) \tag{2-1}$$

Since $n_{cladding}$ is smaller than n_{core}, Snell's law requires that $\cos(\theta_2)$ is larger than $\cos(\theta_1)$. This in turn implies that θ_2 is smaller than θ_1. (The cosine function increases as the angle decreases.) We can also observe that for a sufficiently small value of θ_1 equation (2-1) cannot be satisfied unless θ_2 is an imaginary angle. This occurs because for angles θ_1 smaller than a certain critical angle the solution of Maxwell's equations reveals that there is no field propagating away from the interface in the cladding region. Instead there is only an incident and a reflected wave in the core region. This critical angle (in radians) is obtained by using Snell's law to find the smallest value of θ_1 which results in a real value for θ_2. The largest value that $\cos(\theta_2)$ can take on (for real θ_2) is unity. This implies $\theta_2 = 0$ radians. Solving for θ_1, we obtain

$$\theta_{1\ minimum} \triangleq \theta_{critical} = \cos^{-1}(n_{cladding}/n_{core}) \tag{2-2}$$

A typical value for $(n_{cladding}/n_{core})$ is 0.99. Where this typical value comes from will be discussed later in Section 2.5. If we use 0.99 in equation (2-2) we find that the critical angle is 0.142 rads, or 8.11°. Thus rays traveling at an angle less than 8.11° relative to the reference axis in Figure 2-1 will be totally internally reflected, and guided by the fiber. Higher-angle rays will enter the cladding and be lost due to high levels of scattering and absorption.

2.1.2 Fiber Geometry

It is useful at this point to describe the physical size of the fibers which are typically used in the applications we shall talk about in Part 2 of this book. A fiber consists of a core region and a cladding region (see Figure 2-3). The core region is typically made of silica glass, and typically has a diameter of between 5 μm (0.0002 inch) and 100 μm (0.004 inch). A fiber with a small core (called a single-mode fiber) propagates light in a somewhat different fashion than a fiber with a large core (called a multimode fiber), as we shall describe in Section 2.1.4 below. It is reasonable to ask: small or large compared to what? The answer is: small or large

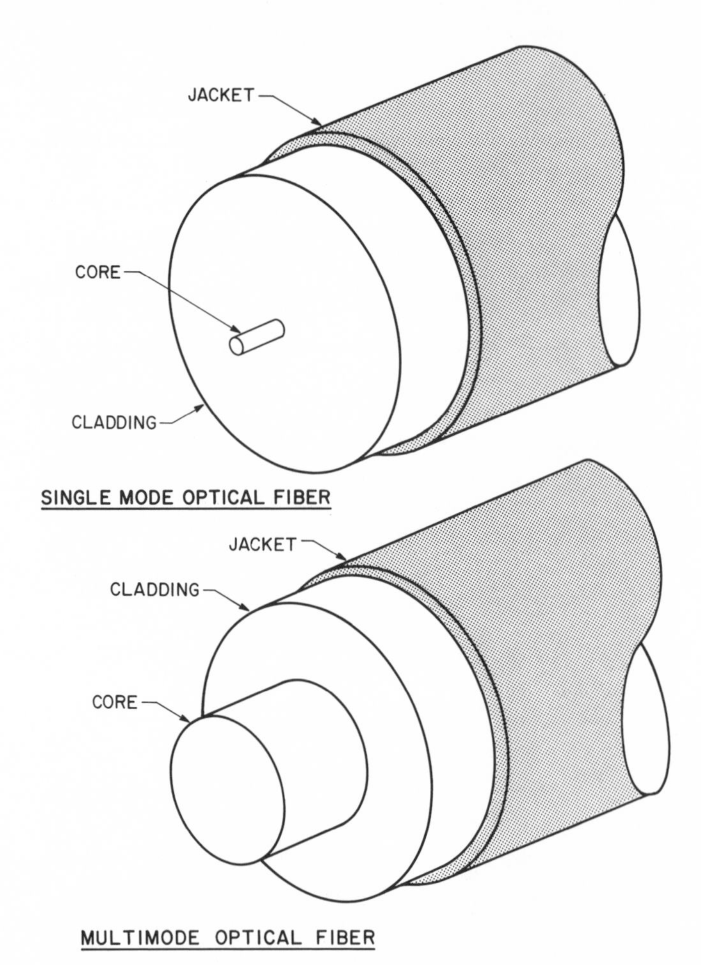

Figure 2-3 Illustration of Fiber Geometry.

compared to the wavelength of the light which will be propagated in the fiber. The wavelength of light for typical optical fiber systems will be between 0.8 and 1.6 μm, with some short-distance links operating at somewhat shorter wavelengths. [Visible light extends from 0.4 μm (violet) to 0.7 μm (red).] A small core fiber has a core diameter which is only a few times as large as the light wavelength. A large core fiber has a core diameter which is tens or even a hundred times as large as the light wavelength. Again, this will affect the propagation characteristics, as will be described below. The core of the fiber is surrounded by a silica glass cladding having a slightly lower optical density (index of refraction), typically between a fraction of 1% and few percent lower. The diameter of the cladding is typically 125 μm (0.005 inches), although it can be larger for some applications. This diameter corresponds to the size of a typical human hair. (Some fibers for special applications are made with a silica glass core and a plastic cladding.) Light tends to be guided in the core of the fiber as described in Section 2.1.1 above. The cladding of the fiber is typically surrounded by a protective jacket. The jacket prevents the fiber surface from being abraded (scratched) and also cushions the fiber from mechanical forces. The size and composition of the jacket depends upon the application, but it is typically one or more layers of plastic material.

2.1.3 Attenuation

If we consider the input–output characteristics of the fiber then, for most applications, attenuation is by far the most important. Attenuation refers to the loss of light energy as a pulse of light propagates down a fiber as illustrated in Figure 2-4. There are two mechanisms by which this loss of energy can occur. In absorption, the propagating light interacts with impurities in the silica glass (or with the silica glass itself at some wavelengths) to cause electrons to undergo transitions. Later these electrons give up this absorbed energy by emiting light at other wavelengths or in the form of mechanical vibration (heat). Thus, in absorption, energy is removed from the propagating pulse and given up later in some other form. The other mechanism for energy loss is scattering. Here geometrical imperfections in the fiber (on a scale which can be small or large compared to the wavelength) cause light to be redirected out of the fiber. Thus in

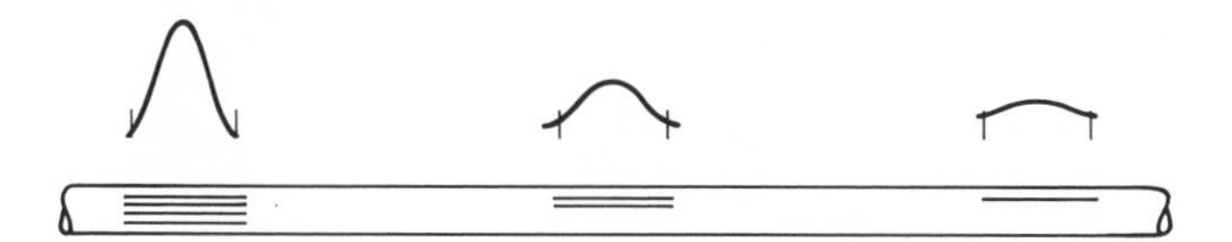

Figure 2-4 Illustration of Attenuation.

scattering the propagating energy leaves the fiber at the same wavelength at which it arrives at the geometrical imperfections.

In the early days of research on the use of fibers for telecommunications (1966–1972) attenuation in optical fibers was dominated by absorption by impurities. Figure 2-5 shows a table of various types of metallic impurities which are commonly found in silica glasses. A calculation was made (in those early days) of the permissible levels that could be present in the silica glass, for each of these impurities individually, while still retaining an attenuation of 10 dB/km (not all that good a value by today's standards, but much better than what was available at the time the table was first made). The "tolerable" concentrations are shown in the figure. One can see that, in some cases, impurity levels of parts per billion were the maximum tolerable. At the time, these impurity levels were difficult to verify by chemical analysis and were very difficult to maintain as raw materials were processed into fibers. Figure 2-6 shows a graph of attenuation achieved vs. time at the wavelength 0.82 μm. This wavelength was used in first generation optical fiber systems because it is compatible with the wavelengths emitted by GaAlAs light sources and the sensitive wavelength range of silicon detectors. In the late 1960s, attenuations were measured in hundreds of decibels per kilometer. (Airports define visibility as 20 dB of attenuation. Thus if you could see at 0.82 μm wavelength, which is just beyond the visible red, and if the loss of a fiber were 100 dB/km, then you could see through 200 m of that fiber.) There was at first a gradual reduction in fiber losses attributed to careful purification of starting materials and careful processing techniques, until 1970, when Corning Glass Works announced a fiber with an astoundingly low (at the time) loss of 20 dB/km [2]. We shall discuss in Section 2.2

ELEMENT	CONCENTRATION ppb
IRON	20
COPPER	50
CHROMIUM	20
COBALT	2
MANGANESE	100
NICKEL	20
VANADIUM	100

Figure 2-5 Table of Absorbing Impurities and Allowable Concentrations for 10 dB/km Loss.

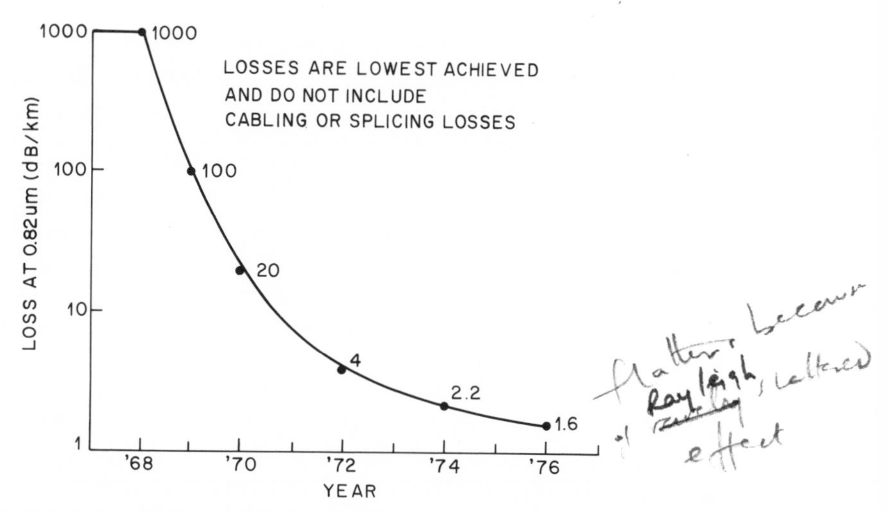

Figure 2-6 Best Reported Attenuation vs Time at 0.82 μm Wavelength.

how they achieved this breakthrough. In 1972 Corning announced a fiber with a loss of 4 dB/km, which demonstrated that long-distance fiber optic communication links were scientifically feasible, and also verified the early predictions [1966] of C. Kao and co-workers [1].

By 1976 the best losses obtained at this wavelength were at around 1.6 dB/km. At this point attenuation was no longer dominated by absorption from impurities in the glass, but rather by a scattering mechanism called Rayleigh scattering. As we shall discuss in Section 2.2 silica glass is by definition not a crystalline material. The atoms within the material are arranged in a somewhat (but not completely) random fashion where the number of atoms in a randomly selected small volume of space is not fixed. As light propagates through the glass, electrons in the glass structure interact weakly with the light by absorbing and reradiating it at the same wavelength (but out of phase). Most of this reradiated light adds to the original propagating light wave, causing only a phase delay (which manifests itself as the index of refraction of the material). However, because of the somewhat random arrangement in space of the atoms of the glass, some of the reradiated light leaves the fiber as what we call scatter. This scattered light, due to the Rayleigh scattering mechanism described above, represents a fundamental lower limit on the attenuation of a particular glass material at a particular wavelength. In the case of silica glass and 0.82 μm wavelength, this lower limit is around 1.6 dB/km. It can be shown that Rayleigh scattering in a particular material varies inversely with the fourth power of wavelength. In fact this is why the sky is blue. The sky scatters light from the sun by a similar mechanism. Shorter-wavelength light (blue light) is scattered more strongly. Thus if we were to use light at a longer wavelength than 0.82 μm we could obtain lower losses.

It is for this reason that around 1975 activities in fiber research (and ultimately development) shifted toward the wavelength range from 1 to 1.6 μm. Of course, one needed light emitters and light detectors which were compatible with this wavelength; but given those, one could obtain much lower values of attenuation [3]. Figure 2-7 shows a graph of the Rayleigh scattering limit in the longer-wavelength range and also the actual loss of an excellent single-mode optical fiber fabricated to operate at that wavelength range. The loss at 1.3 μm is below 0.5 dB/km. The loss at 1.55 μm is below 0.2 dB/km. (The best reported fiber loss to date is 0.16 dB/km at 1.55 μm wavelength.) One can note the loss peak at 1.4 μm wavelength caused by the presence of OH radical (water), which has a strongly absorbing resonance at that wavelength; and which has other resonances also shown. In this fiber, great effort has been expended to remove OH from the material so as to minimize this peak. It is because of the inevitable presence of some OH in the fiber core that fiber systems are typically designed to avoid the use of wavelengths near 1.4 μm. Beyond 1.6 μm the fiber material (silica glass) itself begins to interact strongly with the propagating light, and thus becomes an absorber. If one wished to obtain even lower losses than 0.16 dB/km one would be forced to a different material system. Research on nonsilica glasses has already shown promising results. Losses below 20 dB/km have been reported in these

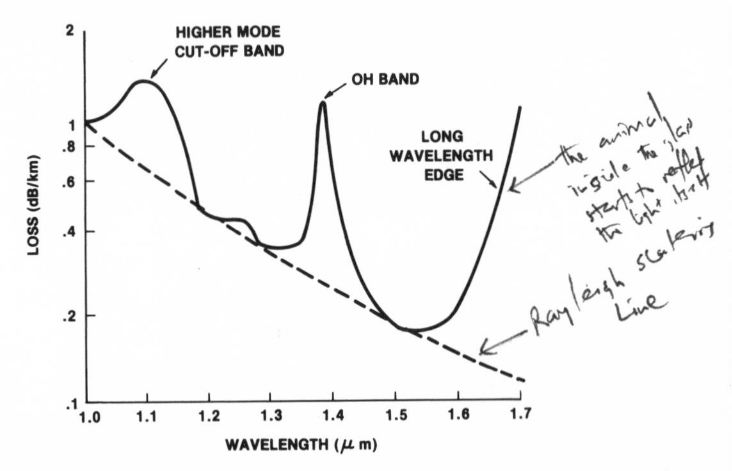

Figure 2-7 Rayleigh Scattering Limit at Long Wavelengths, and Loss vs Wavelength of an Excellent Optical Fiber.

exotic materials. Theory predicts losses below 0.01 dB/km. Whether this can be demonstrated and made practical remains to be seen. The above discussion was concerned with loss mechanisms associated with the fiber itself. In Section 2.3 we shall discuss losses which can be caused in the cabling process (microbending losses) and how these excess losses are controlled.

2.1.4 Delay Distortion

When a pulse of optical energy propagates along a fiber it can spread out in time as shown in Figure 2-8. This phenomenon, called delay distortion, can cause initially separated pulses to overlap in time, making them difficult or impossible to individually distinguish. In the literature delay distortion is sometimes called dispersion; but in this book, in keeping with emerging glossary standards, we shall call the general effect of temporal spreading delay distortion. There are two major reasons why pulses can spread in time. The first effect is associated with the fact that a fiber can typically guide light in a range of angles as shown in Figure 2-9. From geometry, it is apparent that a ray traveling along the axis will reach a given distance in the z direction more quickly than a ray traveling at the critical (maximum) angle relative to the axis. Mathematically, if the speed of light in free space is c (m/s) and if the index of refraction of the core is n_c, then the delay to travel a distance L (m) is Ln_c/c, along the axis. For a ray traveling at the maximum angle, the delay is (from geometry) $(Ln_c/c)\ (\cos\theta_{max})^{-1}$. On the other hand, the maximum angle, θ_{max}, is given from Snell's law by $\cos\theta_{max} = n/n_c$, where n is the index of refraction of the cladding. Thus the maximum delay is $(Ln_c/c)\ (n_c/n)$. The difference in the delays of the fastest and slowest rays is $\Delta t = (Ln_c/c)\ [(n_c - n)/n]$. The first term in this product is the delay along the axis; and is equal to 5 μs/km. The second term in the fractional index difference between the core and the cladding, which is typically around 0.01 (1%) for a standard fiber with a 50 μm core diameter. This means that the delay difference between the fastest and the slowest ray is about 50 ns/km for this type of fiber. This may not sound like a lot;

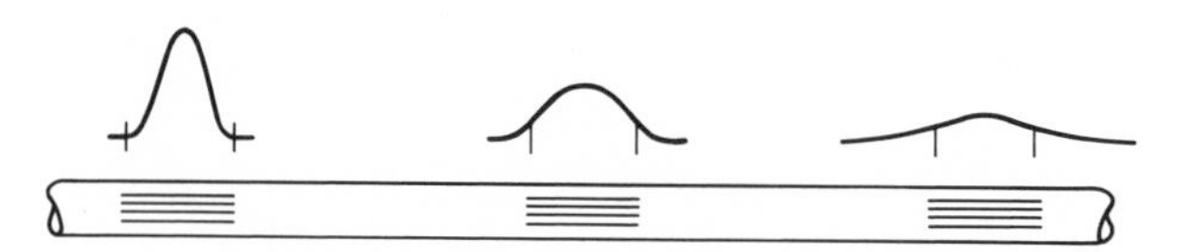

Figure 2-8 Illustration of Delay Distortion.

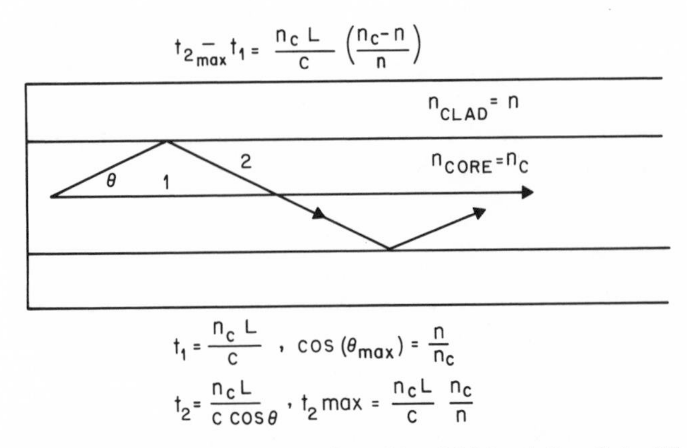

Figure 2-9 Origin of Modal Delay Spread in a Multimode Step Index Fiber.

but if one were to transmit pulses at a rate of 20 million per second, then the pulses would be separated initially in time by 50 ns. According to the above calculation, after only 1 km of transmission, they would be completely overlapping, assuming that all guided angles are initially launched by the light emitter. (Even if all guided angles are not initially launched, mode mixing effects typically couple light to all angles after a few hundred meters of propagation in a cabled fiber.)

There are two ways to reduce this pulse spreading effect (which is called modal delay spreading). One approach is to use a so-called graded index fiber. This type of fiber has a core whose index of refraction (optical density) is highest on the axis of the fiber, and which tapers off roughly parabolically toward the core–cladding interface as shown in Figure 2-10.

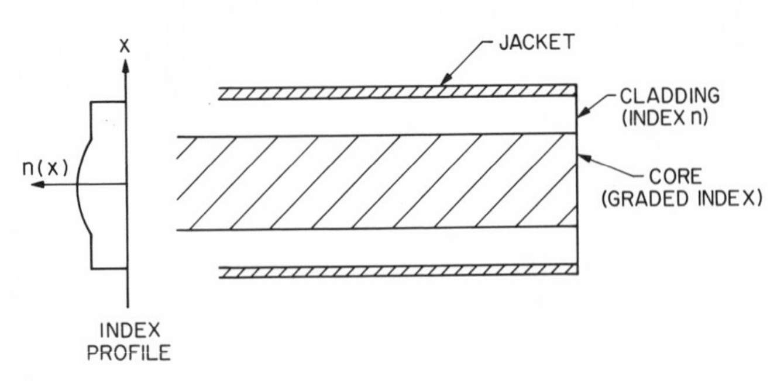

Figure 2-10 Illustration of a Graded Index Fiber.

It can be shown that in such a fiber, rays do not travel the zig-zag paths as indicated in Figure 2-9, but rather helical paths (sinusoidal in two dimensions) as shown in Figure 2-11. At first it would appear that one would still have equal problems with modal delay spreading. Clearly ray 1 shown in Figure 2-11 travels a shorter path than ray 2, in order to propagate a given distance in the axial direction. However, one must recall that the optical density is decreasing as one moves away from the axis. Thus ray 2 can make up in speed (light travels faster in a less dense medium) what it loses in distance. By carefully selecting the profile of the index of refraction (value as a function of distance from the axis) one can arrange for all of the guided rays to propagate at nearly the same axial speed. How well can one do? In theory with very careful control of the index of refraction profile one can improve the pulse spreading by a factor of nearly 1000 in a fiber with a 1% difference between the value of the index of refraction at the axis and the value at the core–cladding interface. In practice one can obtain an improvement of between 20 and 100 in manufactured fibers, depending upon the degree of control exercised and upon whether or not premium quality fibers are selected from a larger lot. This means that instead of the 50 ns/km of pulse spreading one obtains in a step index fiber (with a 1% index step) one can obtain fibers with 0.5–2.5 ns/km of pulse spreading. With 1 ns/km of pulse spreading a 10-km fiber could propagate pulses spaced by 20 ns without severe overlap. Thus a digital stream modulated at 50 Mb/s could be transmitted through 10 km of fiber without problems due to delay distortion. Fabrication of graded index fibers will be discussed in Section 2.2.

There is an alternative to the graded index fiber for controlling pulse spreading. One can use a fiber with a very small core diameter. From our previous discussions it is not obvious why the core diameter would affect pulse spreading. This is because our previous discussions are based on the geometrical optics model of the fiber, where any ray can propagate provided it is at an angle relative to the axis which is smaller than the critical angle. If one examines Maxwell's equations (physical optics) one finds that in fact only certain discrete angles can propagate, corresponding to the discrete set of guided modes in the fiber (which is a dielectric waveguide). These allowed rays are separated in angle of propagation by the amount λ/D (radians) where λ is the optical wavelength in the medium and D is the diameter of the fiber core. Using this result we can modify our geometrical optics model accordingly. Let us consider first the case of a large core fiber having a core diameter of 50 μm. Let us assume that the free space optical wavelength is 1 μm, corresponding to a wavelength in the silica glass of 0.66 μm. The ratio of the fiber core diameter to the wavelength in the material is about 75. This means that the spacing between allowed rays is about 1/75 rad or about 0.8°. If the critical angle

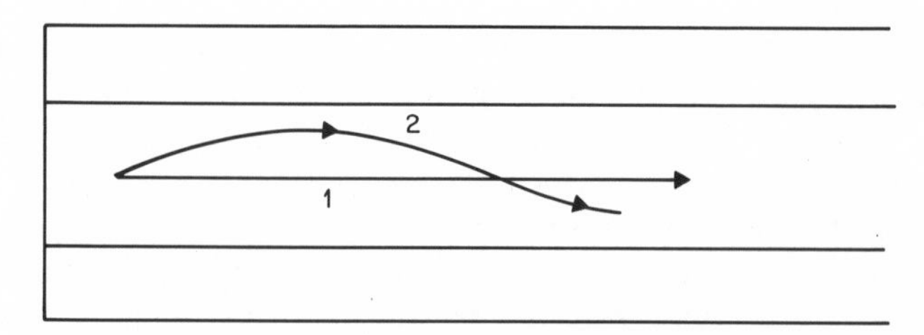

Figure 2-11 Ray Model of Light Propagation in a Graded Index Fiber.

relative to the axis is around 8° (for a 1% index step), then there are 10 allowed rays in each direction away from the axis, or 20 rays in two dimensions. In a three-dimensional fiber there are about 20 × 20 = 400 allowed rays. Thus our original geometrical optics approximation is not all that bad for a large core fiber. Now consider a fiber whose core diameter is 6.6 μm. The ratio of λ/D is now 0.66/6.6 or about 1/10. The spacing between allowed rays is now 1/10 rad or about 5.7°. If the index of refraction step between the core and the cladding is about 0.5% then the first allowed ray away from the axis (5.7°) will be above the critical angle. In this case, only the axial ray can propagate. If we have only one ray, then we eliminate the delay distortion associated with different rays propagating at different speeds. A fiber which can only propagate one ray is called a single mode fiber because in the physical optics model, there is only one guided solution to Maxwell's equations (i.e., one mode). A single mode fiber is more difficult to couple light into and more difficult to splice. However, because of its ability to carry very narrowly spaced pulses (high bandwidth) it is becoming increasingly popular in many applications.

In addition to pulse spreading due to different modes traveling at different speeds, one has to be concerned with different optical wavelengths traveling at different speeds. A typical light emitter will emit over a range of wavelengths. This range can be very narrow (almost zero in some circumstances) for a laser emitter, but it can be quite broad (more than 5% of nominal) for incoherent light emitters. Silica glass has an index of refraction which is wavelength sensitive. This is why a prism can spread white light into its colors. One can show that the propagation speed through a material depends upon its index of refraction and upon the first derivative of the index of refraction with respect to wavelength. The propagation speed will vary with wavelength if the second derivative of the index of refraction with respect to wavelength is not zero. This effect, called material dispersion, can be characterized by the variation in delay (ns/km) with respect to wavelength change (nm) and thus has the units of ns/km-nm. Figure 2-12 shows the material dispersion effect in silica glass over the range of wavelengths of interest in typical fiber optic systems. For

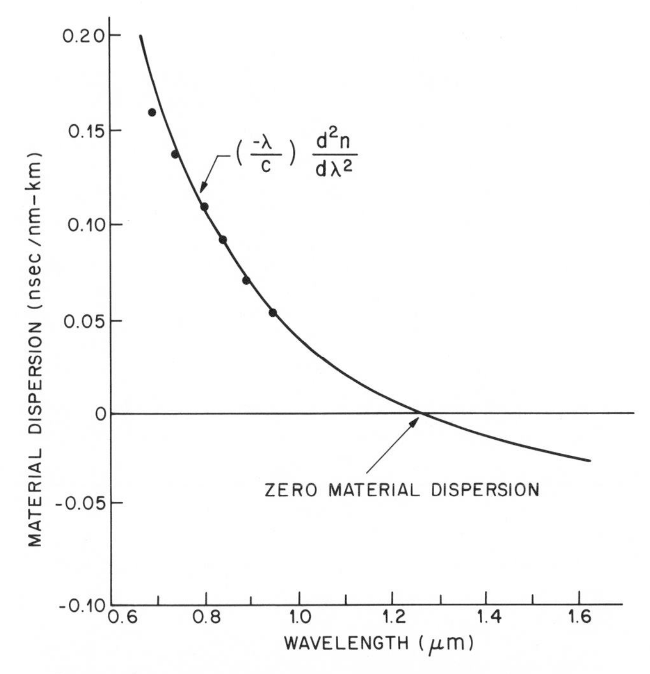

Figure 2-12 Material Dispersion vs Wavelength in Silica Glass.

example, at 0.82 μm wavelength the material dispersion effect is about 0.1 ns/km-nm. This means that an incoherent light-emitting diode emitter with a nominal wavelength of 0.82 μm and an optical bandwidth of 50 nm will produce a pulse spreading (due to different wavelengths traveling at different speeds) of 0.1 × 50 = 5 ns/km. This pulse spreading is larger than that caused by modal delay spreading in a graded index fiber.

To reduce this effect one could use a coherent optical emitter (laser) having a much narrower spectral width. Alternatively one could operate with light at a wavelength where the material dispersion is lower. Looking at Figure 2-12 we observe that the material dispersion effect passes through zero at about 1.3 μm. This is fortuitously a wavelength where the optical attenuation can be quite low (0.5 dB/km). For this reason 1.3 μm is a very popular wavelength for fiber system operation.

As a further note on the mechanisms which cause pulse spreading we must mention that in single mode fibers waveguide propagation effects can cause a wavelength dependence of the single mode propagation constant which is similar to material dispersion. In fact, it can have a sign which is opposite to that of material dispersion, with the effect that the frequency of zero net dispersion is shifted. For complicated single mode fiber designs,

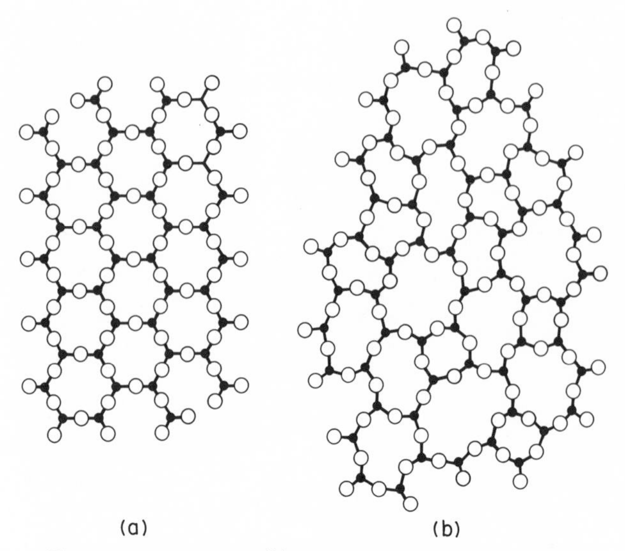

Figure 2-13 Crystalline (a) and Glassy (b) Silica Structures.

one can also achieve a wide range of wavelengths where the dispersion effect is uniformly very low. This opens up the possibility of using the fiber simultaneously over a range of wavelengths while still avoiding dispersion.

It should also be pointed out here that although waveguide dispersion can cancel out material dispersion, dispersion cannot cancel out modal delay spread. It can be shown that these two pulse spreading effects add as the "sum of the squares." That is, the total delay distortion from modal delay spread and dispersion is obtained by squaring the individual effects (in ns/km of spreading), adding the squares, and finally taking the square root of the result.

2.2 Glass and Fiber Making

The purpose of the sections below on glass and fiber making is to familiarize the reader with the material which forms the heart of systems using fiber optics technology and with the methods by which this material is fabricated into useful fibers.

2.2.1 The Nature of Glass

When we think of glass most of us think of the material from which windows and drinking glasses is made. In actuality glass is (technically speaking) not a particular material, but rather a state of matter such as solid, liquid, or gas. In particular, a glass is a solid state of matter in

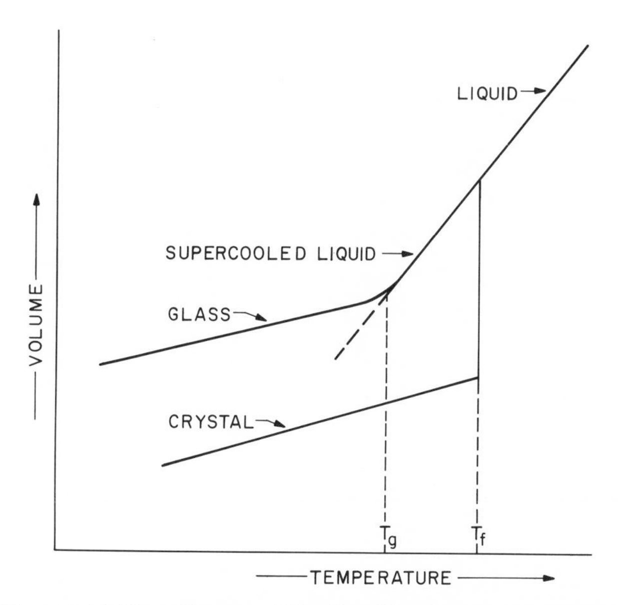

Figure 2-14 Phase Diagram of Liquid and Solid States of a Material.

which the atoms are not in a regular array (as in a crystal), but where the interatomic spacings and bond angles are irregular. Figure 2-13 illustrates the compound silicon dioxide in both its crystalline state (left) and its glassy state (right). If cooled down slowly, a liquid will from a crystalline solid at a temperature called the freezing temperature, as shown on the graph in Figure 2-14. This is an abrupt (as a function of temperature) process accompanied by a change in volume at the transition. However, if a liquid is cooled down very rapidly, then at a temperature below the freezing temperature, called the glass-forming temperature, the liquid may gradually solidify in the glassy state as shown in Figure 2-13b. Figure 2-15 gives a table of various materials which can form glasses along with their glass-forming temperatures. For example water, which normally freezes at 273 K, will form a glass if rapidly cooled to 140 K from the liquid state.

The particular material of interest for making fibers is silicon dioxide mixed with oxides of germanium, boron, phosphorus, and other elements to adjust its optical density slightly from that of pure silicon dioxide glass (silica).

2.2.2 Making Glass Preforms

In the early days of optical fiber research, scientists attempted to make fibers from what were called soda–lime–silicate glasses. These so-called mixed glasses contain sodium and calcium atoms which act as glass

SUBSTANCE	T_g (K)
NATURAL RUBBER	200
POLYVINYL CHLORIDE	347
H_2O	140
HPO_3 + 12.5 % H_2O	359
GLYCEROL	183
GLUCOSE	305
n – PROPANOL	95
$Ca(NO_3)$ + 62% KNO_3	315
SELENIUM	303
$ZnCl_2$	373
BeF_2	570
GeO_2	800
SiO_2	1350

Figure 2-15 Materials Which Have Glassy States.

modifiers to further disrupt the structure (break bonds) of the silicon dioxide. (See Figure 2-16.) This results in a glassy material which melts at a lower temperature than pure silica. It was believed that this lower melting temperature would be necessary in order to make fiber fabrication practical. In particular it was felt that since pure silica melted at a very high temperature, it was inevitable that impurities from the container holding the molten glass during the fabrication process would contaminate the melt. This turned out not to be an insurmountable limitation, as will be described shortly. In those early days, fiber losses were high because it was difficult to obtain pure starting reagents (sodium carbonate, calcium oxide, and silicon dioxide powders); it was difficult to maintain the purity as these were mixed and melted into a mixed glass; and it was difficult to obtain uniformity in the melt. Losses of soda–lime silicate glasses eventually were reduced below 10 dB/km at 0.85 μm wavelength; but before that occurred, activity switched to so-called high silica glasses.

In 1970 Corning Glass Works demonstrated a method for forming low loss (below 20 dB/km) fibers from mixtures of silicon dioxide and oxides of germanium, boron, and other atoms [2]. In their process they avoided what others were concerned with, namely, melting the high silica glass inside of a container. Briefly their process is the following (see Figure 2-17). They start out with very pure silicon tetrachloride, a volatile liquid which can be purified of metallic contaminants to below 1 ppb with

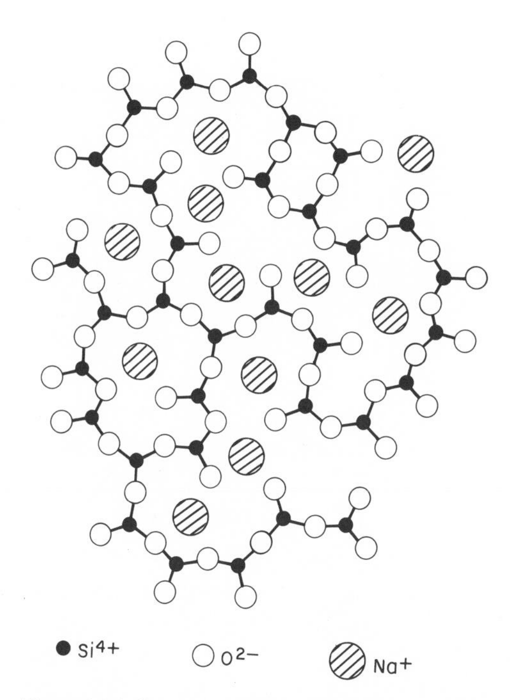

Figure 2-16 Structure of Soda–Lime–Silicate Glass.

proper processing. Through this liquid they bubble pure oxygen, which picks up some of the $SiCl_4$. The mixture is burned in hydrogen to form tiny particles of very pure SiO_2. These particles, formed in the burner, are directed at a rotating target mandrel where they deposit in the form of a soot (fine particles). As a mandrel rotates and the burner translates back and forth over the mandrel a soot "preform" is built up which is typically about a meter long and perhaps ten centimeters in diameter. As the preform is built up the composition of the deposited soot particles can be varied. For example, germanium tetrachloride and boron trichloride can be added to the oxygen stream to raise or lower the refractive index of the deposited material. When a sufficiently large soot preform has been deposited, the preform is placed in a furnace, which causes the soot particles to melt together into a clear solid preform. This process is called sintering. Finally the target mandrel is removed, and the preform is again heated, causing it to collapse into a solid rod. It is from this solid rod that a fiber can be drawn, as will be described below. Corning called their process the chemical vapor deposition process. It is often referred to as the outside chemical vapor deposition (OCVD) process.

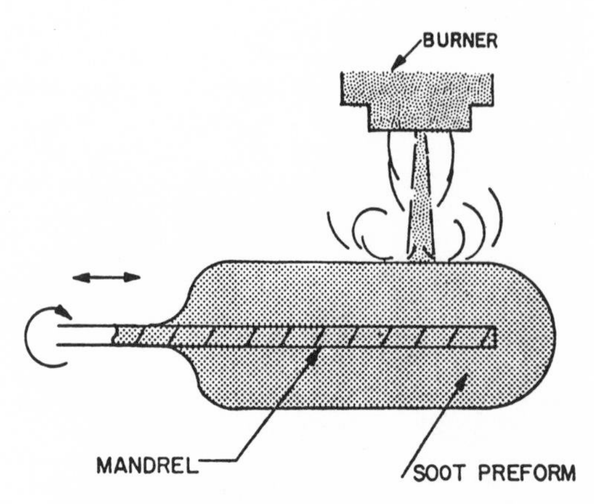

Figure 2-17 CVD Glass Preform Fabrication.

Following Corning's announcement, another process was disclosed by Bell Laboratories which is called the modified chemical vapor deposition process or the inside chemical vapor deposition process. This is illustrated in Figure 2-18. Pure oxygen is bubbled through silicon tetrachloride, germanium tetrachloride, and mixed with other materials, and allowed to flow through a starting tube of high quality silica (fabricated by some other process). The tube is held in a glass working lathe containing two synchronous chucks. The synchronous chucks rotate the tube while allowing access to the ends. Where the cutting tool would be placed in a metal working lathe one places a torch which translates back and forth across the rotating tube. As gases flow past the torch, they ignite to form particles of pure glass (soot particles) which deposit on the walls of the tube downstream from the flame. As the flame passes over the soot particles it consolidates them into a solid clear glass. In this way layer upon layer (possibly of varying composition) of new glass can be deposited inside the tube. The inner cladding layers are deposited first, followed by what will become the fiber core. After sufficient new glass has been deposited inside the tube, the gases are shut off, the temperature of the torch is raised, and the tube collapses into a solid preform rod. A typical tube is about 1 m long and 25 mm in diameter.

Recently, a third process was disclosed by NTT, in cooperation with Sumitomo, Furukawa, and Fujikura companies in Japan, which is referred to as the vapor axial deposition process. Here, as shown in Figure 2-19, material is deposited on the end of growing preform. The resulting soot preform is sintered into a solid rod, as in the outside chemical vapor deposition process.

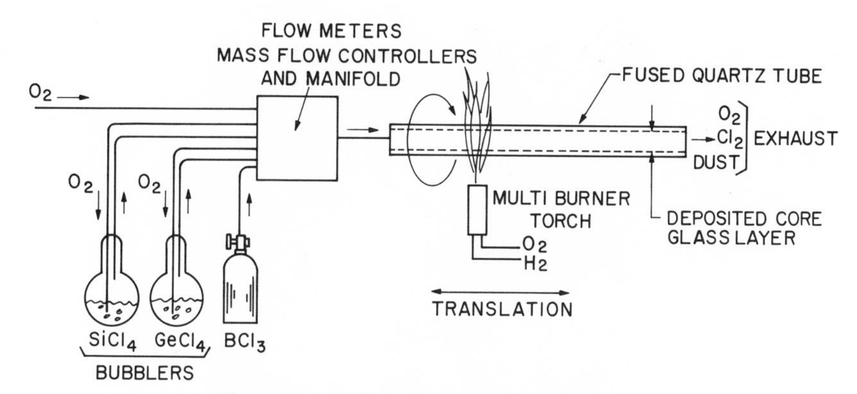

Figure 2-18 MCVD Glass Preform Fabrication.

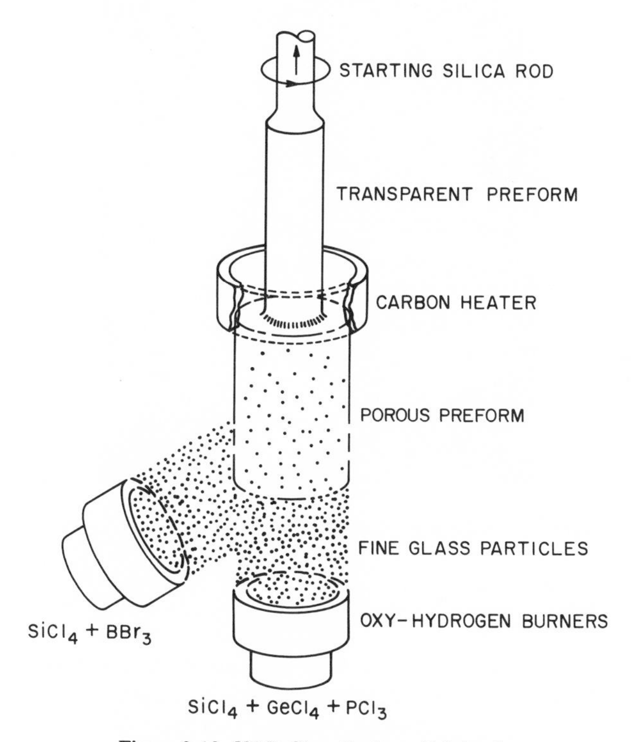

Figure 2-19 VAD Glass Preform Fabrication.

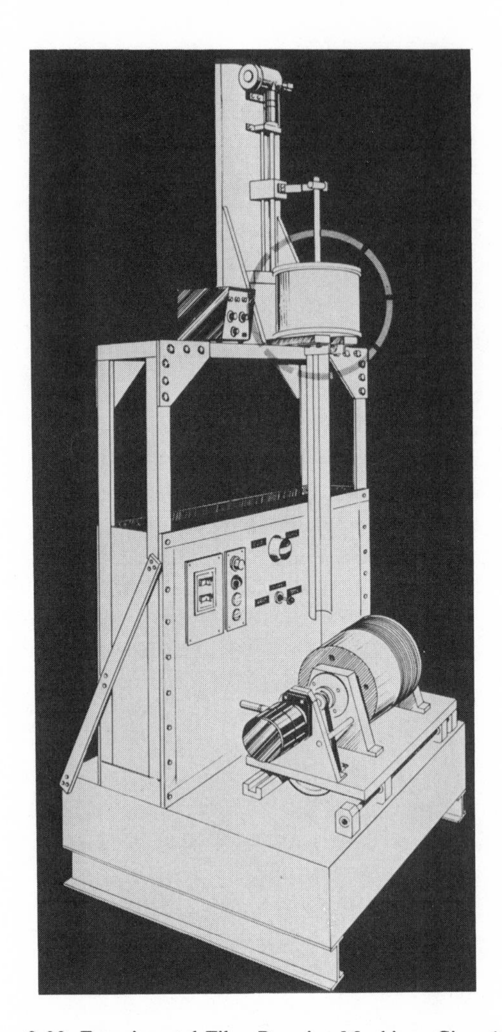

Figure 2-20 Experimental Fiber Drawing Machine. Circa 1970.

Each of these processes has been used to make low loss high quality single and multimode fibers. Each can provide adequate control to accurately form the core and cladding regions, including a graded index core region. Each process is constantly being improved to increase the rate of deposition of materials, the efficiency of use of starting materials, and the yield of the fabricated fibers. These factors determine cost, which ultimately may favor one process over another.

2.2.3 Drawing Fibers

Having fabricated a clear silica glass preform by any of the methods described above, one can now proceed to draw a fiber. Figure 2-20 shows an artist's concept of a simple fiber-drawing machine (a real version of which was actually used in early research at Bell Laboratories). At the top of the drawing tower (circled) is a cylindrical oven, which may be heated either by electrical resistance elements or burning gases. The preform rod is lowered into the oven at a controlled rate, where it reaches a sufficiently high temperature to soften.

To start the draw a pair of tongs is used to pull the softened preform end down (like taffy) to a thin strand (fiber), which is then attached with tape to a rotating take-up drum. At this point, a fiber can be drawn whose diameter depends on the rate of rotation of the drum and the rate of feed of the preform into the oven. In today's production environment, the fiber drawing machine is similar to the relatively simple machine described above, but with certain important embellishments. It is necessary to carefully control the fiber outside diameter to facilitate splicing and cabling. To accomplish this an optical instrument (which works on diffraction effects) is placed below the furnace to measure the fiber diameter as it is drawn. The results of this measurement are electrically fed back to the controllers for the take-up drum and the preform feed motors. Thus the diameter is controlled with a closed loop control system. Drawing speeds can be in excess of 5 m/s; and the response time of the closed loop is a factor in determining the spatial correlation function of diameter variations. It is important to note that the ratio of surface area to volume in a given length of fiber is much larger than in the preform. Thus the fiber cools very quickly as it leaves the heating zone. This allows for the application of protective coatings (first jacket layers) using coating dies placed just beyond the diameter measuring instrument. These coatings protect the fiber from abrasion, which would weaken it considerably, as will be described further in Section 2.3 below. Finally, before winding the fiber on to the drum, each piece of drawn fiber passes over a pair of pulleys which apply a fixed amount of strain (stretching per unit length). This "proof-test" is used to verify that all of the fiber is strong enough to use in actual cables.

2.3 Cabling

In order for fibers to be used in practical applications they must be placed in cables to protect the fibers from the environment and to facilitate handling and splicing. Environmental factors of interest include tensile

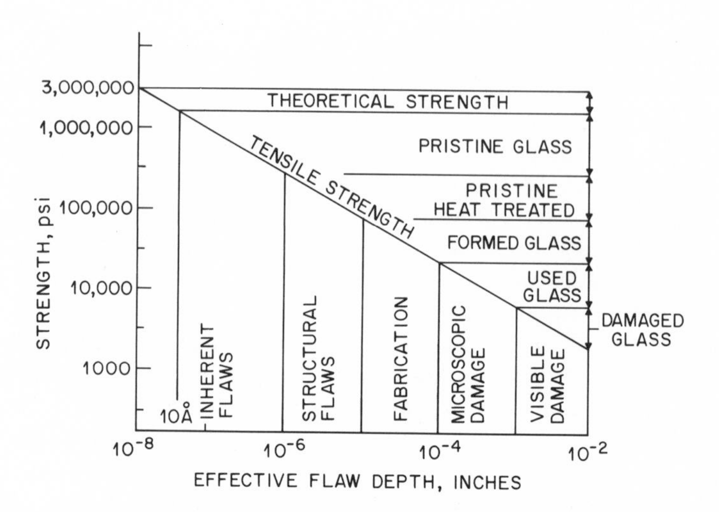

Figure 2-21 Glass Strength vs Flaw Depth Calculated for a Wedge Shaped Scratch (Flaw) in the Surface of a Plate of Glass Under Tension Perpendicular to the Direction of the Scratch.

stress, crushing, bending, corrosives, fire, humidity, and rough surfaces (which cause the fiber loss to increase owing to a phenomenon known as microbending, which will be described below).

Fibers can be nonintuitively strong provided that the surface of the fiber is free from flaws. Figure 2-21 shows a graph of glass strength in psi vs flaw depth in the glass surface. It is well known that the way to break a piece of glass is to scratch it along the line of the desired break. Similarly it can be seen from this figure that the short term breaking strength of fibers is very sensitive to the degree of perfection of the fiber surface. The strength of a fiber can equivalently be characterized by its breaking strain. A flawless surface allows a strain of several percent without breakage. Typically fibers are proof-tested in manufacture (as described in Section 2.2 above) to guarantee a minimum breaking strain before cabling. Typical proof-test strains vary from 0.5% to 1%, depending upon the cable requirements.

Theory and experiment have shown that a fiber which can support a given short term strain may break in time without any increase in tension due to a mechanism called stress corrosion or static fatigue. In the presence of strain and humidity (OH) small flaws in the fiber surface will grow in time via a chemical reaction between the OH radical and the silica. Theory and experiment indicate that this static fatigue mechanism proceeds at a rate which is very sensitive to the ratio of long term applied strain to short term breaking strain. Thus by designing the cable and cable installation procedures in such a way as to guarantee that the long term strain imposed on the fibers is a small fraction of the proof-test strain (e.g., 20%) one can anticipate a fiber lifetime (due to stress corrosion) of several decades (even in the presence of 100% humidity). Thus although some cable installations have been engineered to keep the fibers in a low humidity environment this may not be necessary if the in-place residual strain is properly controlled.

To protect the fibers from excessive tensile strain, the cable includes strength members which distribute the tensile load over a sufficiently large area to keep the tensile stress within safe bounds. These tensile strength members can be made of steel or nonmetallic yarns designed for this purpose (Kevlar for example). To protect the fiber from crushing forces stiff materials can be fabricated into the outside layers of the cable (e.g., stranded steel wires or metallic tubes) and/or the fibers can be recessed into various cushioning layers. Excessively tight bends can strain the fibers beyond allowable limits. Stiffness to limit the bending radius and to prevent accidental kinking of the cable must be engineered into the cable structure. Additionally fibers are often wound in helical paths within the cable to allow for reasonable cable bending radii. The outside layer of the cable must be resistant to corrosives which can be anticipated in the particular applications and which can attack either the glass fibers or the cabling material itself.

When fiber cables are placed within buildings local ordinances may require that the cable material be flame retardant and additionally that fumes released from the cable in the presence of fire meet certain restrictions.

When low loss fibers were first manufactured a mechanism called microbending loss was observed. When fibers were wound on drums for shipment or storage their losses increased from low values to very large values. This was explained by mode coupling theory. The fibers were being forced to conform to the mechanically rough surface of the drum which imposed bends in the direction of the fiber axis having mechanical periods in the range of several millimeters. Even though these bends were very small in mechanical amplitude, calculations showed that they were sufficient to cause modes (rays) in the fiber to couple amongst each other and to the continuum of unguided radiation modes. In order to reduce this effect to tolerable values it is necessary to mechanically isolate the fibers from bends having periods in the range of millimeters to centimeters. This is accomplished by fabricating a combination of buffer layers of varying mechanical properties around the fiber and by careful design of the means by which the fiber is routed inside the cable. The expansion coefficients of

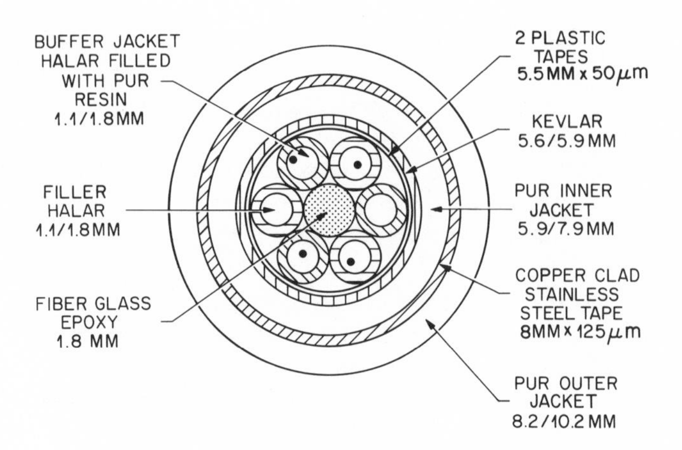

Figure 2-22 Loose Tube Cable Structure Example.

the cabling materials relative to the fiber are also an important factor in determining whether microbending losses will increase when the cable is exposed to changing temperatures.

When a large number of fibers are contained within the cable structure it is important to provide markings for identifying individual fibers and to organize the fibers in ways which facilitate splicing. Some cable designs place fibers in ribbons or other orderly structures which allow for multiple fiber splicing with proper fixtures (each fiber is a separate transmission path, but several fibers are spliced at once). Figure 2-22

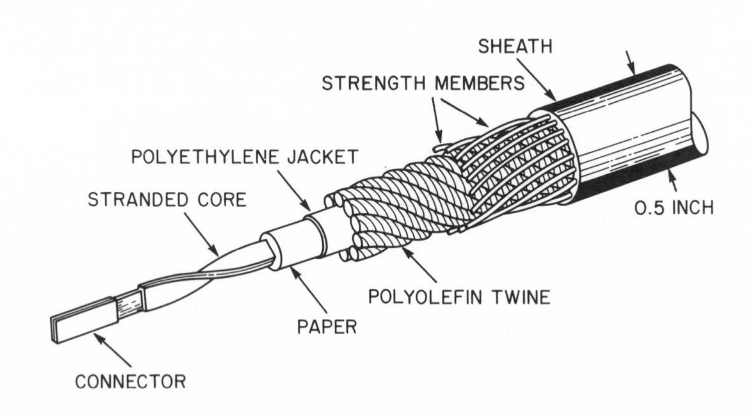

Figure 2-23 Ribbon Cable Structure Example.

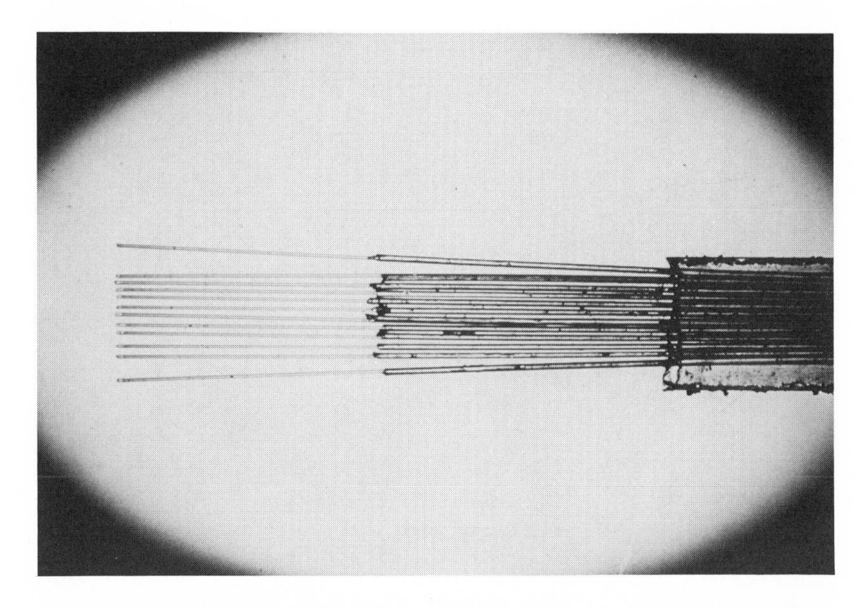

Figure 2-24 Photo of a 12-Fiber Ribbon. Reprinted by Permission of Bell Communications Research and AT&T Bell Laboratories.

shows an example of a cable containing four fibers which are held in loosely fitting tubes. Such an approach is referred to as a loosely buffered fiber cable design. The buffered fibers are stranded into the relatively large cable structure using machines which are similar to those used to make metallic cables. Tensile strength is provided by a carbon loaded yarn called Kevlar which has a very high elastic modulus (ratio of stress to strain). Buckling resistance is provided by a fiberglass core. Except for the stainless steel rodent shield, this cable design is entirely dielectric (no metallic conductors). A version without this shield would be suitable for installation in environments where lightning or high voltage differences are present. Figure 2-23 shows a design where the fibers are held in tightly fitting ribbons. This is referred to as a tightly buffered fiber cable design. Each ribbon in this design contains twelve fibers; and up to twelve ribbons can be stacked to form a 144-fiber array. In Section 2.4 below we shall show how a mechanical splicing method can be used to splice all of the fibers in an array of ribbons simultaneously. Thus this cable not only protects the fibers, but also organizes them for easy identification and splicing. Figure 2-24 shows a ribbon of twelve fibers. Starting at the left we can see the uncoated glass fibers approximately 125 μm (0.005 inch) in diameter. We can see the plastic coating which is approximately 250 μm (0.010 inch) in diameter. Twelve coated fibers side by side form a ribbon approx-

Figure 2-25 Photo of a 12-Ribbon Cable. Reprinted by Permission of Bell Communications Research and AT&T Bell Laboratories.

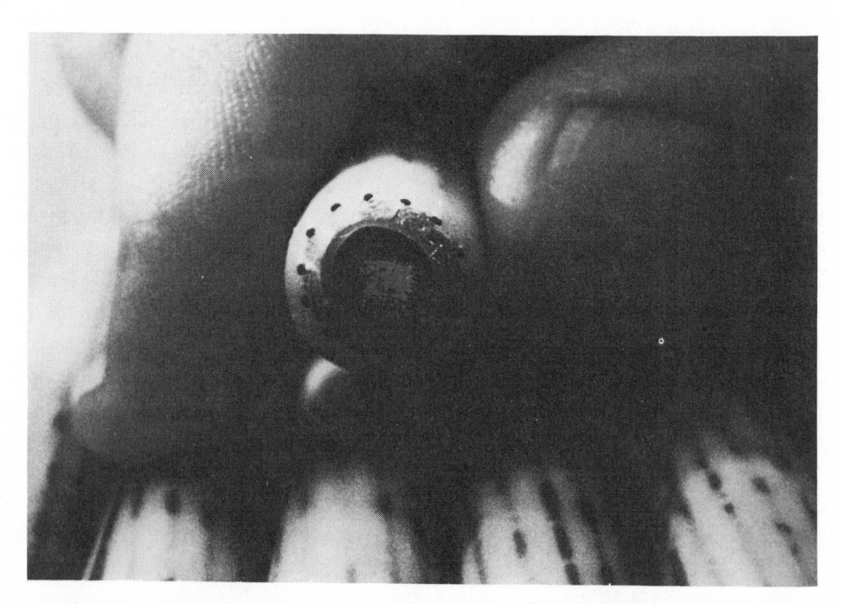

Figure 2-26 Photo of the End of a 12 Ribbon Cable. Reprinted by Permission of Bell Communications Research and AT&T Bell Laboratories.

imately 3 mm (0.120 inch) across. The adhesive tapes which hold the fibers together are also visible. Figure 2-25 shows the end of a cable with twelve numbered and color coded ribbons exposed. Figure 2-26 shows the end of the cable with several fibers illuminated by visible wavelength (red) light. The steel strength members are visible, as is the core of twelve ribbons. The twelve ribbons in this photograph are actually fitted with a 12 × 12 fiber array splicing fixture which holds the 144 fibers in a precisely defined matrix. This will be further described in Section 2.4 below. Fiber cables have been designed for a wide range of applications ranging from fiber guided missiles, to undersea cable, to internal equipment wiring, to directly buried telecommunications cables. Each application has its own requirements and tradeoffs. The common goals are to protect the fibers from excessive strains and to prevent microbending loss.

2.4 Splicing and Connectorization

In practical applications it is necessary to interconnect fibers in sections of cable and to join fibers, attached to sources and detectors, to cable sections. A low loss connection between two fibers can be implemented by aligning the cores and allowing guided light to leave one core and propagate into the next. The amount of light which is not coupled (and consequently lost) depends upon the mechanical accuracy of the alignment, the physical match between the fiber parameters (core size, index of refraction profile), and reflections at the interface due to the presence of air or other materials in the small gap between the fibers.

In this section and the remainder of this book we shall refer to a connection which is intended to be permanent as a splice. We shall refer to a connection which is intended to be rearrangeable as a demountable connection implemented with a connector.

To understand the exact loss experienced at a given splice or connector it is necessary to know the distribution of the light power amongst the guided modes (rays) of the fiber, the field patterns associated with the modes, and other physical details. However, a good deal of insight and some excellent approximations for the loss of a connection can be obtained from some simple geometrical arguments.

Figure 2-27 shows a splice between two identical multimode (large core compared to the wavelength of the light) fibers where certain alignment defects are illustrated. These include lateral offset, longitudinal offset, and angular offset. When identical fibers are offset laterally a loss occurs which can be approximated by the degree of overlap of the core areas. In this model light rays which leave the incoming fiber will be lost if they are not incident upon the core of the outgoing fiber. Angular offsets

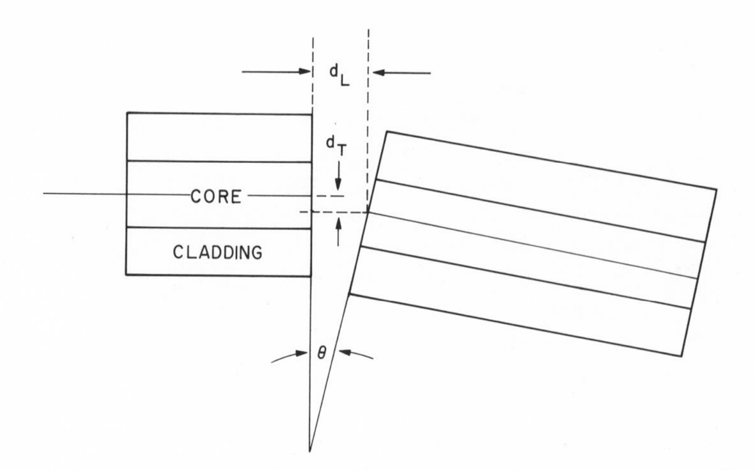

Figure 2-27 Illustration of Splicing Alignment Imperfections.

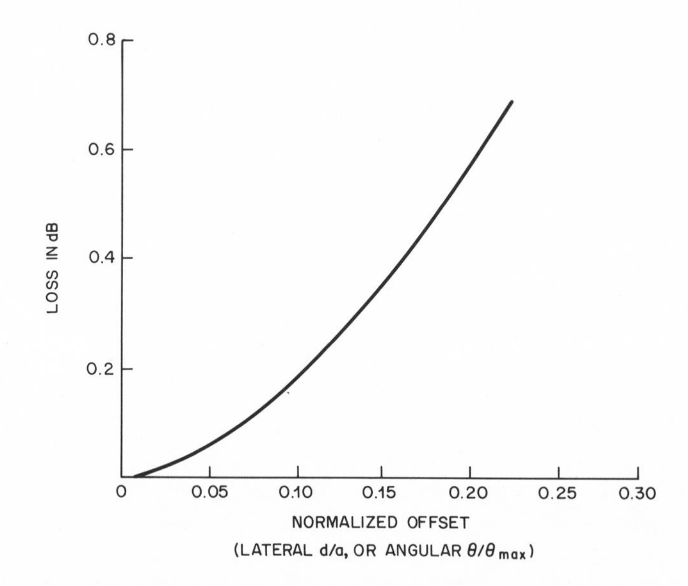

Figure 2-28 Multimode Fiber Splice Loss vs Lateral and Angular Offsets.

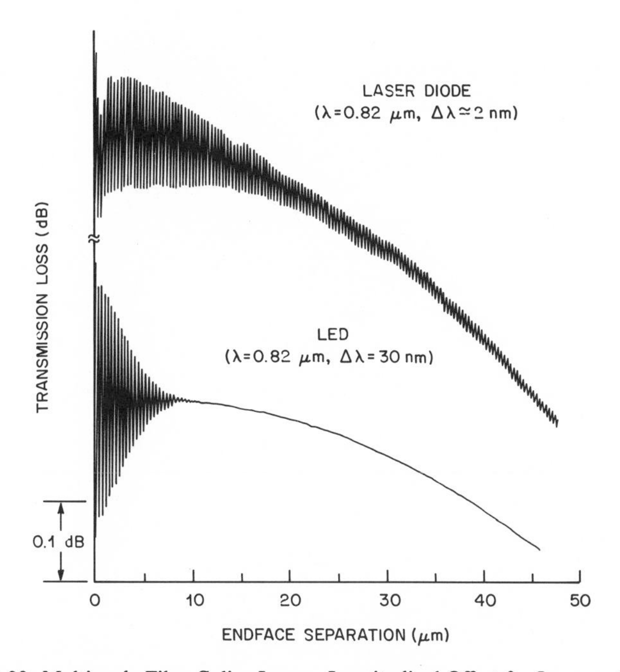

Figure 2-29 Multimode Fiber Splice Loss vs Longitudinal Offset for Laser and LED Excitation.

result in losses which can be attributed to rays which are within the critical angle in the incoming fiber but beyond the critical angle in the outgoing fiber. Longitudinal offsets result in losses which can be attributed to the expansion of the beam (diffraction) as it enters the gap between the incoming and outgoing fibers, and to reflections at the two interfaces between the fiber ends and the intervening material. The reflection at an interface between silica glass and air is 4% per interface. This reflection can be reduced by placing a material in the gap which has an index of refraction close to that of silica glass (1.5).

Figure 2-28 shows calculated losses for multimode fiber due to lateral or angular offsets, where d/a is the ratio of lateral offset to core diameter and θ/θ_{max} is the ratio of angular offset to critical angle. Figure 2-29 shows measured loss vs longitudinal offset, including resonance effects from reflections at the fiber to air interfaces, not including the reflection losses. Figure 2-30 shows the calculated losses due to offsets for identical single mode fibers, not including reflections. Losses due to mechanical alignment errors or reflections, with nominally identical fibers, are referred to as extrinsic losses.

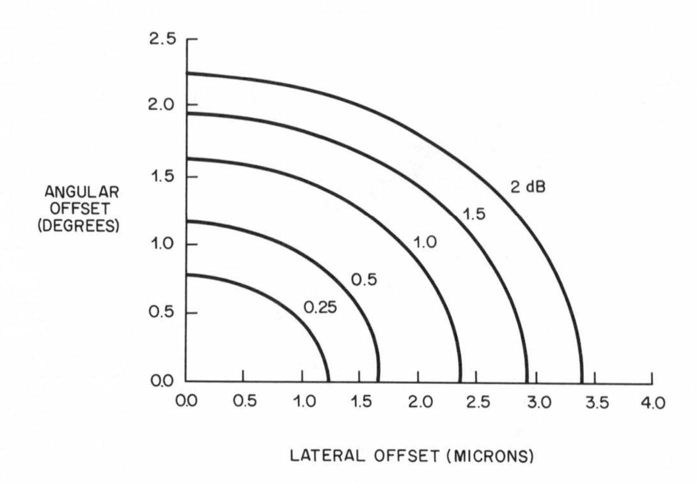

Figure 2-30 Single Mode Fiber Splice Loss vs Lateral and Angular Offsets.

Additional losses (beyond extrinsic losses) can occur if the fiber parameters are mismatched. For example fibers with different core diameters cannot couple in a lossless fashion in both directions since light filling the larger core cannot be coupled into the smaller core. Similarly fibers with nonidentical index of refraction profiles will not couple perfectly.

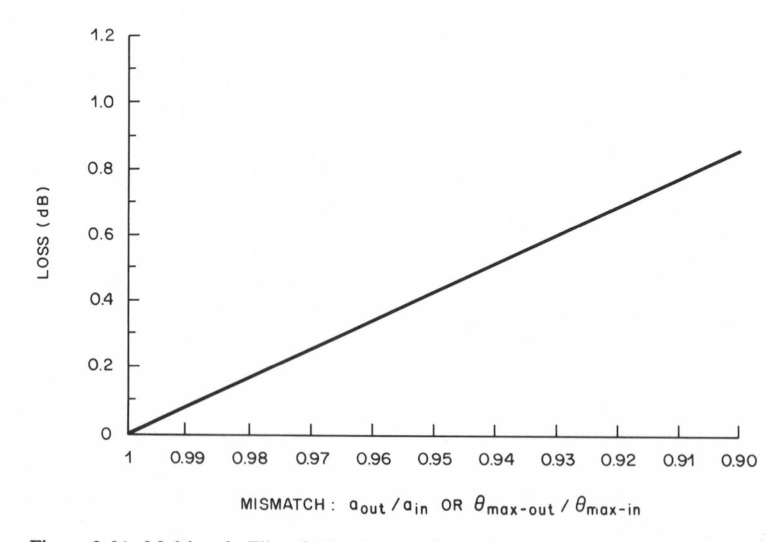

Figure 2-31 Multimode Fiber Splice Loss vs Core Size or Critical Angle Mismatch.

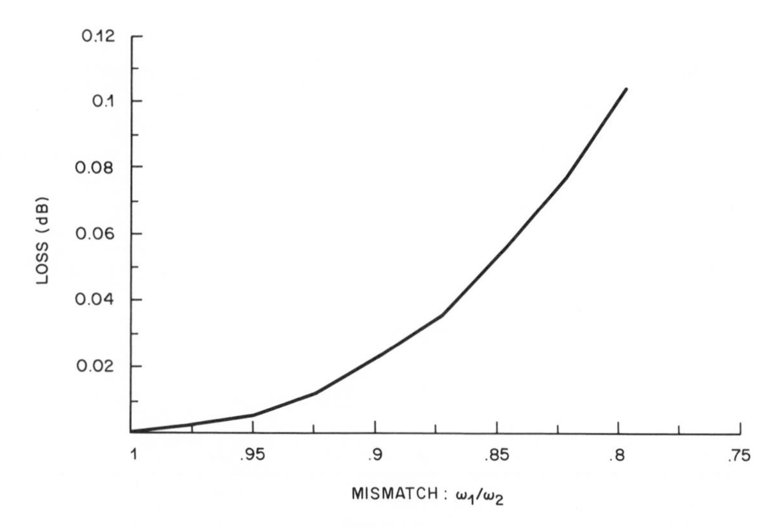

Figure 2-32 Single Mode Fiber Splice Loss vs Spot Size Mismatch.

In single mode fibers mismatch can be best characterized by a parameter known as the spot size. This corresponds to the half power width of the field pattern supported by the single mode fiber at a given wavelength. To a very good approximation the loss of a perfectly aligned splice can be predicted from the mismatch of this parameter.

Figure 2-31 shows a curve of loss vs core size or critical angle mismatch for perfectly aligned multimode fiber cores (in the propagation direction of larger-to-smaller fiber). Figure 2-32 shows a curve for splice loss vs. mismatch of the single mode field pattern spot size in single mode fibers for perfectly aligned cores. Losses resulting from fiber parameter mismatches are called intrinsic losses.

In order to achieve low splicing losses with practical splicing techniques great care is exercised in fiber manufacture to maintain well controlled fiber parameters (cladding diameter, core diameter, index profile, single mode fiber spot size) and good concentricity of the core within the cladding.

There are two popular methods of splicing fibers. These are the V-groove mechanical alignment method and the fusion splicing method. Figure 2-33 illustrates the V-groove method by an example. A capillary tube with a square cross section is fabricated by fusing four solid rods together, and drawing this relatively large structure down to a physically small capillary tube. The tube is cut to a length of several centimeters, and the ends are funneled out. Fibers are inserted into each end of the tube and posi-

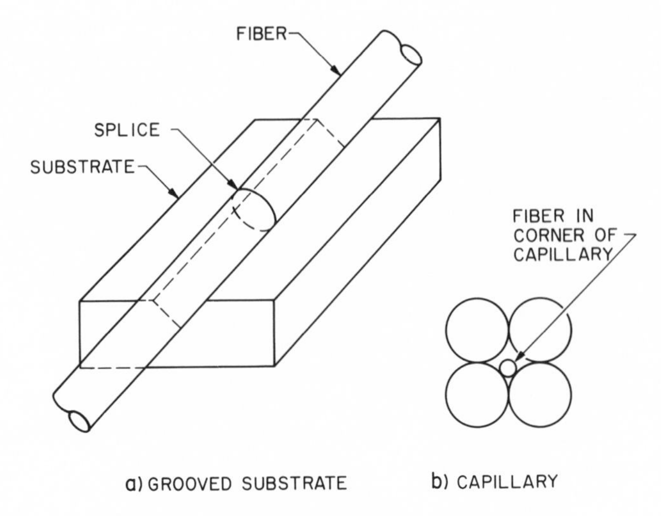

Figure 2-33 V-Groove Splicing Method Examples: (a) Grooved Substrate, (b) Square Capillary.

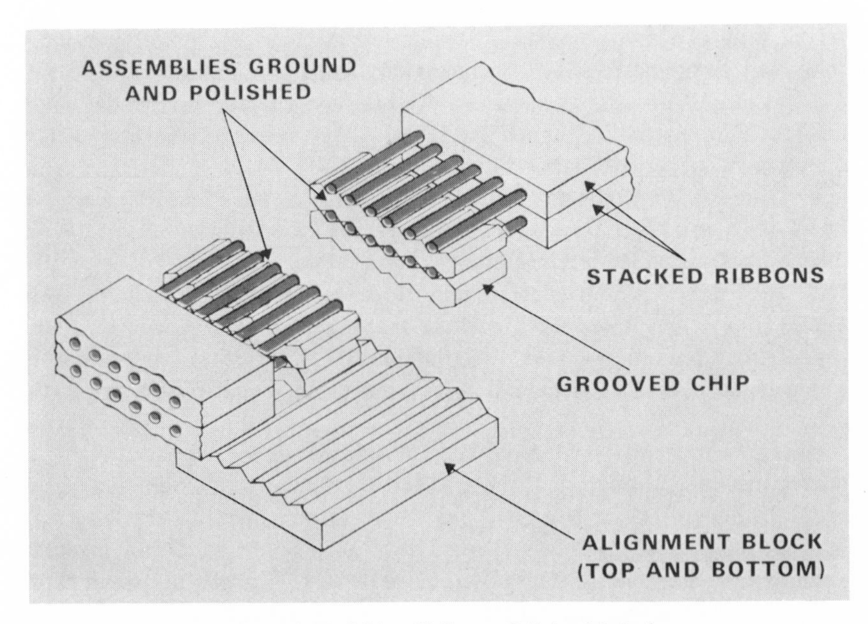

Figure 2-34 Ribbon V-Groove Splicing Method.

tioned into one corner of the square cross section. They are pushed into the tube until they meet. If the fibers have flat and perpendicular ends and if the cores are centered in the claddings, a very good alignment can be made. The fibers are fixed in place with epoxy.

Figure 2-34 illustrates another example of the V-groove splice used with ribbon cables. The ribbons are stripped of all plastic coatings and the remaining glass fibers are laid into grooves formed by etching techniques in silica chips. The positions and sizes of the grooves are very well controlled. Layers of ribbons are built up into a matrix (array) by stacking chips as shown. When the array is completed, the pieces are fixed in place with epoxy or other material. The array is then scribed and cleaved (broken) to form a reasonably flat and square end surface, which is further polished to optical flatness. Two arrays can be joined to form a multifiber splice using an alignment chip as shown. With this technique up to 144 multimode fibers have been spliced simultaneously with individual splicing losses below 0.25 dB. Reflections are minimized by placing an index matching material in the gap between the arrays. The entire spliced structure is held together with a mechanical spring.

In the multifiber ribbon array splice, the fiber ends were cleaved and polished. In individual fiber splices good flat ends can be formed on fibers by careful cleaving techniques. The fibers are scribed and placed under tension (often over a curved surface). This causes the scratch formed by the scribe to propagate into the fiber to form a flat perpendicular surface on the fiber end. Figure 2-35 shows the ends of fibers which have been prepared by cleaving methods. If not done properly one obtains the defects shown in (a) and (b). These defects prevent the fibers from being placed in proper mechanical alignment. However, with proper technique, the flat end shown in example (c) can be obtained. In this figure, the fibers were bent over a curved surface after scribing and during the application of breaking tension. The parameter R signifies the radius of curvature of the bend applied.

For single mode fibers the perpendicularity of the surface formed on the fiber end relative to the fiber axis is critical for obtaining a very low loss splice (see Fig. 2-30). Since single-mode fibers have very low losses at long wavelengths (below 0.5 dB/km) one typically strives for splice losses below 0.1 dB. To achieve this, perpendicularity within less than a degree is important.

In the fusion splicing method, fibers are aligned using a mechanical fixture equipped with a small microscope. Micropositioners are used to physically move the fibers into position under visual inspection. Once aligned, an arc is drawn across the interface to fuse the fibers together. Fusion eliminates the reflections at the interface and if done properly

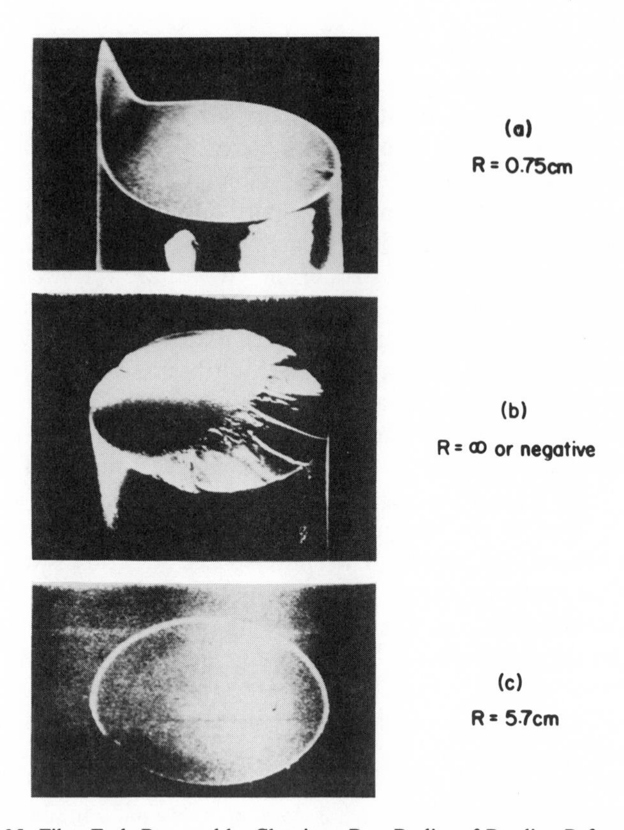

Figure 2-35 Fiber Ends Prepared by Cleaving. R = Radius of Bending Before Tension.

results in a strong permanent connection. (It is important in fusion splicing and V-groove splicing not to damage the unprotected fiber surfaces, which would make them susceptible to breakage under relatively small strains.) In some sophisticated fusion splicing apparatus automated means are used to align the fiber cores once they have been approximately aligned manually. This is achieved by transmitting light through the fibers being spliced during the splicing operation and monitoring the throughput as a measure of alignment.

At this time it is not clear whether V-groove splicing or fusion splicing is superior.

A connector performs the same functions as a splice, namely, physical alignment of the fiber ends being connected. However, a connector has the additional requirement for assembly and disassembly in a quick and convenient manner. Typical connector designs consist of a cylindrical object formed around the fiber in such a manner as to have the fiber end up with its core very accurately concentric with a reference surface of the cylinder. Two cylindrical connector halves are mated in a matching sleeve. Figure 2-36 illustrates an example of a popular connector invented by P. Runge of AT&T Bell Laboratories. A bullet-shaped plug (cylindrical object) is injection molded around the fiber in such a way that the fiber is very accurately concentric with the tapered end surface. As the molding material cures there is some shrinkage, but the shrinkage is uniform enough to maintain the concentricity of the fiber and the taper. Further, the angle of the taper is preserved in the shrinking process. After curing, the plug is placed in a cone-shaped fixture and the end surface of the plug is polished to a precise length relative to the reference surface (taper). Two tapered plugs fit into a biconical sleeve which has two very concentric tapered sockets back-to-back. The assembly is held together by tightening mechanical hardware into threads molded into the parts. Notable advantages of this connector design are ruggedness (no exposed fiber ends) and abrasion resistance (reference surfaces meet only at the bottom of insertion travel). This connector design also allows for easy cleaning of the parts to remove dust and debris.

Figure 2-37 illustrates another connector concept where precisely aligned lenses are used to collimate (expand) the optical beams at the interface. The advantage of this type of design is that the connection is less susceptible to interruption by a dust particle. Thus this type of connector has been proposed for military tactical applications. Such a connector is less sensitive to lateral offsets but more sensitive to angular offsets.

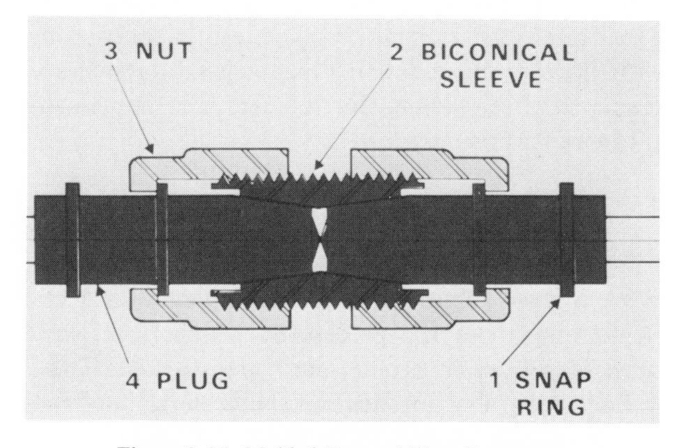

Figure 2-36 Molded Tapered Plug Connector.

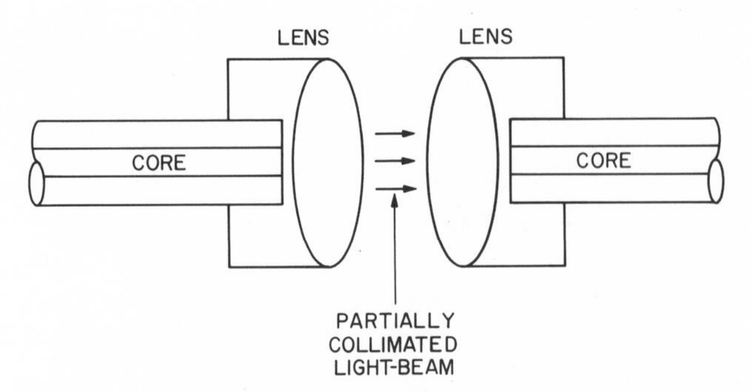

Figure 2-37 Expanded Beam Connector.

For some types of connectors – and particularly for single mode applications – the parts may be actively touched up after assembly by machining or other means while monitoring the core concentricity with the reference surface (facilitated by illumination of the core by accessing the other end of the fiber). For connectors the nominal loss and the repeatability of achieving an acceptable loss value are both important. Depending upon the application, resistance to moisture and vibration may also be important considerations.

Many connector designs are most easily implemented by fabricating the connector in a factory environment onto a short fiber pigtail. The pigtail can then be spliced, in the field, onto a longer fiber. However, field mountable connectors of somewhat reduced performance (insertion loss) also exist.

Although connectors require precision tolerances this does not necessarily imply high manufacturing costs (relative to metallic cable connectors). Cleaver connector designs can be manufactured by automated processes, with little or no manual adjustments or touch-up. Thus with sufficient volume to offset tooling development and implementation, incremental costs can be surprisingly low.

2.5 Selecting Fiber Parameters

Before leaving the subject of fibers, cables, and connections it is useful to say something about the process by which fiber parameters (core diameter, index of refraction profile, etc.) are selected. These parameters impact upon fiber cost, fiber attenuation, pulse spreading, ease of cabling, and connection loss.

For multimode fibers (large core compared to the wavelength of transmission) we see the following tradeoffs.

A larger core makes connecting fibers easier and allows for more light to be coupled into the fiber from a multimode source (e.g., a light-emitting diode). However, a larger core diameter increases the fiber cost (because the core is doped with costly materials like germanium) and increases the susceptibility of the fiber to microbending loss when it is cabled (not obvious from our discussions on microbending above, but discussed in the literature).

"Numerical aperture" (N.A.) is defined as $(n_{core}^2 - n_{cladding}^2)^{1/2}$, which is approximately equal to $n_{cladding}\,(2\Delta)^{1/2}$, where Δ is $(n_{core} - n_{cladding})/\,n_{cladding}$. The N.A. is also the sine of the largest incident angle captured in the fiber core, when the fiber is illuminated from a material of unity refractive index, as shown in Figure 2-38. A higher numerical aperture allows the fiber to support more guided modes. This enhances coupling to multimode sources. However, a larger numerical aperture is achieved by increased doping of the core, which causes increased Rayleigh scattering loss due to compositional fluctuations in the material, and increases cost.

A larger cladding diameter increases the susceptibility of the fiber to strain when it is bent around curves. However, too small a cladding diameter can also make the fiber fragile. Also, the cladding must be thick enough to prevent evanescent fields extending from the core into the cladding from experiencing high losses at the cladding—jacket interface.

In single mode fibers, the core diameter and the numerical aperture must be small enough to assure that only one mode propagates in the wavelength range of interest. However, too small a core may result in a guided mode which extends too far into the cladding and is susceptible to microbending loss.

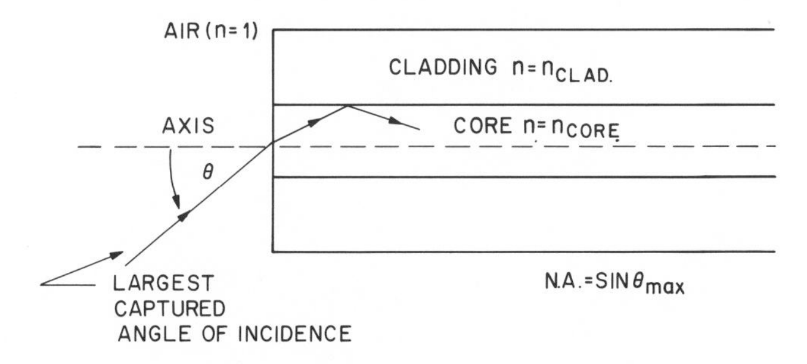

Figure 2-38 Definition of the Fiber Numerical Aperture.

Based on these tradeoffs, fiber and cable designers have been able to agree on at least some standard fiber designs which strike a balance between cost and performance. An example is the 50 μm core, 125 μm cladding, 0.2 N.A., graded index fiber.

Problems

1. What is the maximum amount of fiber (meters) which can be drawn from a preform 1 m long and 2.54 cm (1 inch) in diameter?

2. A fiber has an attenuation rate of 0.5 dB/km. What percentage of the propagating light power is lost (absorbed and/or scattered) in a length of 1 m?

3. A fiber with a step index profile has a core with refractive index 1.50 and a cladding with refractive index 1.49. What is the maximum angle (degrees) that a light ray may have relative to the axis inside the fiber and still be guided by total internal reflection? What is the maximum angle (degrees) a light ray incident upon the fiber core from air outside the fiber may have relative to the fiber axis and still be captured by the fiber? Relate (mathematically) these two maximum angles to the core and cladding indices of refraction for arbitrary values of these indices.

4. How much absolute delay (microseconds) is experienced by a light signal propagating through 10 km of a fiber with core index 1.50?

5. What is the delay difference between the fastest and slowest propagating rays (ns/km) in a step index fiber with core index 1.50 and cladding index 1.49?

6. The material dispersion of a fiber at 0.85 μm wavelength is 0.1 ns/km-nm. How much delay difference (ns/km) is experienced by components of an optical source at wavelengths 0.825 μm and 0.875 μm?

7. What frequency (Hz) corresponds to the optical wavelength 0.85 μm? How much bandwidth (Hz) is occupied by an incoherent optical source with a spectral width of 50 nm and a nominal wavelength of 0.85 μm?

8. The propagation constant for a plane wave in a material (β) is given by $2\pi\, n/\lambda$, where n is the material refractive index and λ is the free space wavelength of the propagating wave. The group delay (τ) is given by the partial derivative of β with respect to ω, where ω is $2\pi f$,

and f is the frequency of the propagating wave (in Hz). Derive expressions for the group delay (τ) and the material dispersion (partial derivative of τ with respect to λ) in terms of: the wavelength, the speed of light in free space and derivatives of the index of refraction with respect to wavelength (note we assume that n is a function of the wavelength (λ) and not a constant).

3

Optical Sources and Transmitters

3.1 LEDs and Lasers

To construct an optical communications system one requires a source of optical power and a means for modulating that source. A suitable source for an optical fiber communications system must have certain characteristics which include the following: emission at a wavelength within a window of low fiber transmission loss, efficient conversion of prime power to light coupled into the fiber, high reliability, ease of modulation, adequate modulation speed capability, sufficient ruggedness, ease of coupling the source output into the fiber, adequately low noise, adequately high linearity of modulation, sufficiently narrow spectral width (range of wavelengths in the emitted light), and other more subtle requirements. No source can provide ideal characteristics, but the requirements above eliminate many candidate sources either because they are totally impractical for the fiber system application or they are clearly inferior to other alternatives. (Obviously all of these requirements must be met at an acceptable cost if the source and the system are to be practical.)

In the early days of fiber optic system research two candidate sources emerged as adequately meeting most or all of the requirements listed above. These are the semiconductor light-emitting diode (LED) and the semiconductor injection laser diode (ILD). (There was a third source which was given some consideration as a possible candidate, but was dropped because it appeared to be more expensive to fabricate and more difficult to modulate. This was the miniature Nd-YAG laser pumped by light-emitting diodes. Because of its long intrinsic time constants, it requires an external modulator for most applications. It is possible that interest in this device will reemerge in the future.)

The LED and ILD are both solid state semiconductor devices which can be fabricated by batch processes (as will be discussed in Section 3.1.2 below). They can be fabricated from various semiconductor material systems, which allows the device designer to select the desired wavelength of emission. In particular, devices fabricated in the gallium–aluminum–arsenide material system can emit in the range of wavelengths between 0.8 and 0.9 μm. Devices fabricated in the indium–gallium–arsenide–phosphide material system can emit in the range of wavelengths between 1.0 and 1.6 μm. Both LEDs and ILDs can be modulated by varying the electrical current used to power the devices (direct modulation). The achievable direct modulation rates range from 20 MHz to beyond 1 GHz for LEDs (depending upon the materials, the device design, and tradeoffs against other parameters), and up to 5–10 GHz for the fastest ILDs. Although the amount of light power coupled into the fiber is typically a small fraction of the drive power, this electrical-to-optical conversion efficiency is adequate for most applications (and is typically much better than what can be obtained with alternative devices). The spectral width of an LED is relatively large, which limits its range of applications. The spectral width of a laser can be very small (a single frequency of emission) depending upon the device design. The impact of laser spectral properties will be discussed in Section 3.1.1 below and also in Chapter 6.

Both LEDs and lasers can have very high reliability, and are compact, mechanically stable, devices. Lasers are susceptible to damage from electrical abuse and are somewhat sensitive to high temperatures, as will be discussed in Section 3.3. In summary, both LEDs and ILDs have found important applications in a wide range of systems.

Figure 3.1 shows a schematic drawing of a typical light-emitting diode. The device consists of a number of layers of semiconductor material, some of which are p-doped and some of which are n-doped as shown. Where the p-doped and n-doped materials come together one has a p-n junction. When holes and electrons are injected into the junction (by applying a current in the forward biased direction) they combine and give up an amount of energy equal to the charge of a electron multiplied by the semiconductor band gap (in volts) between the valence and conduction bands. This energy can be given up as a photon of light or in the form of mechanical lattice vibration (heat).

The simplest type of light-emitting diode would just have two layers — one p-doped and one n-doped. Unfortunately such a simple structure would not be capable of efficient conversion of electrical drive power into light power captured by a fiber. There are several reasons for this. The light emitted by the LED can be reabsorbed before it leaves the device. To

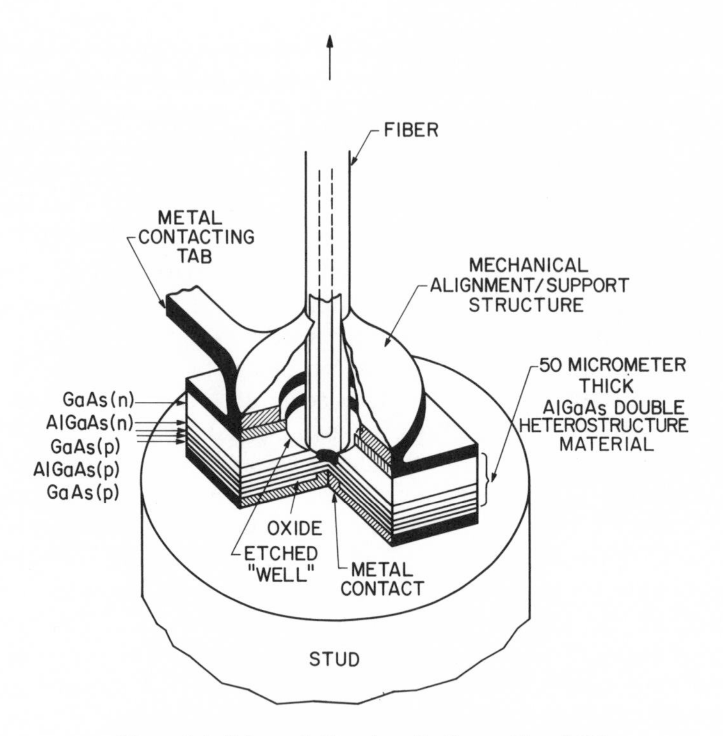

Figure 3-1 Schematic Drawing of a Burrus Type LED.

minimize this problem a well is etched to allow the fiber to be brought close to the junction. However, in a simple two layer device the holes and electrons combine in a relatively thick layer on either side of the junction. Furthermore the process of etching the well too close to the junction can introduce mechanical damage to the material, which results in nonradiative recombination (producing heat instead of light). To alleviate these problems a multilayer structure is used, where the layers are made of semiconductor compounds of varying composition. For example, the active layer shown might be made of pure GaAs material while the layers on either side might be made of GaAlAs where the ratio of gallium to aluminum is 90/10. Note that the aluminum is not a small quantity dopant but is present in substantial percentage. The use of this layered "heterojunction" approach leads to some interesting results. The energy band structures of the layers are different (both the absolute levels and the band gap). As a result potential barriers are formed on either side of the active layer, which confine the holes and electrons to a thin volume within the active layer. Thus all of the photons are created within this thin volume. In addition, the layers containing aluminum are relatively transparent to the light emitted by the active layer. Thus it is not necessary to etch the well all the way down to the active layer (thereby avoiding the nonradiative recombination due to damage). Thus the layered structure allows much more of the light generated within the device to reach the fiber.

The fiber can only capture that portion of the light which illuminates its core. The use of lenses cannot provide coupling from a large light-emitting area to a small fiber core — as will be discussed in Section 3.2 below. Thus it is wasteful of electrical drive power to inject current into the junction over a large area. For this reason a dot contact is defined by an insulating layer to create a column of current of limited area aligned with the well. Finally, to provide efficient heat sinking, the device is often mounted with the substrate up to bring the junction close to the heat sink. The result of all of this will be discussed in Section 3.1.1 below. A device with this structure is called a Burrus type LED after its inventor C.A. Burrus [6].

Figure 3-2 shows the structure of the typical ILD. In this case we again have a series of heterogeneous layers (heterostructure), but the light is emitted from the side of the device. A laser is an oscillator (as opposed to an LED which is a broadband optical noise source). To obtain oscillation one needs gain, feedback, and saturation. Nature will always provide saturation. We must provide the other two items. To obtain gain one injects sufficient current into the device so that a condition called a population inversion exists in the active layer. When holes and electrons combine there is a brief period of time when they are about to emit a photon of

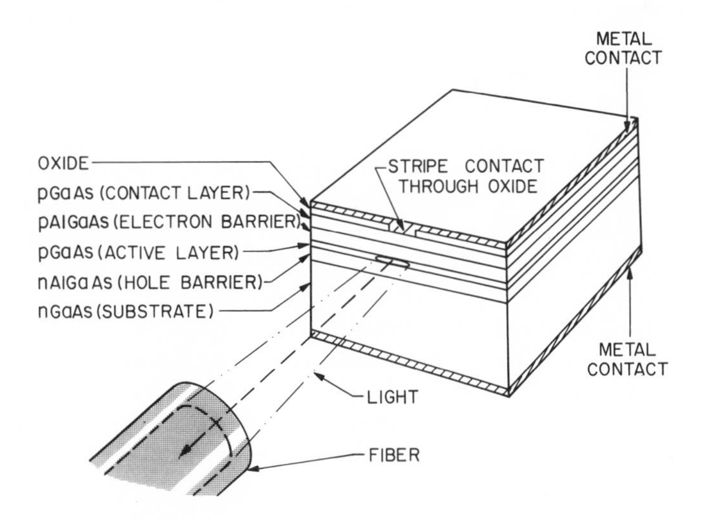

Figure 3-2 Schematic Drawing of a Stripe-Contact Gain-Guided Injection Laser Diode.

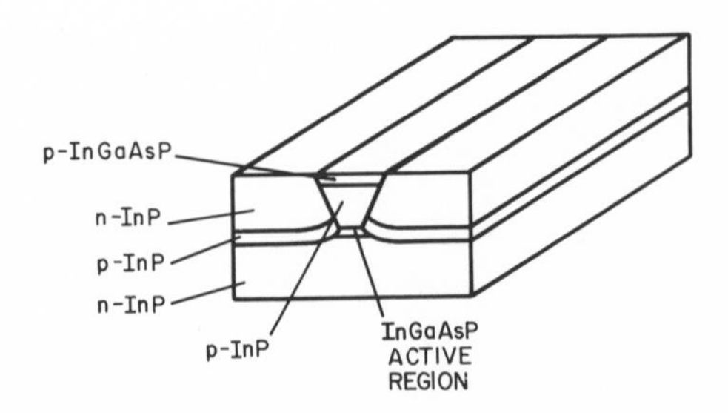

Figure 3-3 Schematic Drawing of an Index-Guided Injection Laser Diode.

light, but have not done so yet. If in that time period an existing photon of light passes by, it can stimulate the hole and electron to add their light energy synchronously to the existing field. Thus the existing field grows in amplitude as it travels through the medium. The reverse process can also occur. That is, an existing photon can be absorbed to produce a hole—electron pair. In order to have gain dominate over loss one must inject a sufficient density of hole—electron pairs into the junction to have the stimulated emission process dominate the absorption process.

To provide feedback an optical cavity is formed as follows. For the direction front-to-back, perpendicular surfaces are cleaved (fabricated by scoring and breaking) onto each end of the device. The interface between the semiconductor material and air produces an approximately 30% reflection. This reflection, when combined with the gain in the active layer, is adequate to result in a unity round-trip gain from any point within the active layer to the front face, across the active layer to the back face, and back. Feedback in the top-to-bottom direction is formed by the layers on either side of the active layer. By good fortune, they have a lower index of refraction and thus the light is guided within the active layer by total internal reflection. Feedback in the left to right direction is provided by one of two mechanisms. If the cross-section of the layers is uniform left-to-right as shown in Figure 3-2, feedback can be obtained by the higher effective refractive index of the region in which current is flowing (defined by a stripe contact on the top of the device). This is called gain guiding. If the layers have a more complex structure (as shown in Figure 3-3), then the layers themselves form a three-dimensional waveguide for guiding the light. This is called index guiding.

In order for the device to operate (achieve gain and lasing) at room temperature, the population inversion must be confined to a small volume of space (to limit the injected current per unit area) [7]. The layers on either side of the active layer not only provide waveguiding for the light, but form potential barriers which confine carriers to a small cross-sectional area.

Thus with this design we obtain carrier confinement, gain at acceptable current densities, and field confinement (feedback)— all of which are needed to produce a practical laser (oscillator).

3.1.1 Input–Output Characteristics

When we apply sufficient current to an LED or ILD light will be emitted. In order to understand the capabilities and limitations of these devices in system applications it is necessary to understand some details regarding their output power vs applied current characteristics, the spectral characteristics of their emitted light, their modulation speed capabilities and sources of limitations, the spatial properties of their emitted light fields, and the effects of temperature and aging on relevant characteristics.

3.1.1.1 Input–Output Characteristics of LEDs

An LED is a noise source which emits its noise in a band of wavelengths centered about its nominal optical wavelength. The nominal wavelength of the device is determined by the nominal energy gap between the semiconductor valence band and conduction band. For gallium–aluminum–arsenide material this band gap is approximately 2×10^{-19} joules (J), corresponding to a photon at a wavelength of approximately 0.85 μm (depending upon the composition of the light-emitting active layer). However, the actual energy difference between a hole in the valence band and an electron in the conduction band can deviate from the band gap energy by about the Boltzmann energy $kT = 4 \times 10^{-21}$ (at room temperature). Thus the actual energy of the emitted photon when a hole in the valence band combines with an electron in the conduction band can vary about its nominal by about $4 \times 10^{-21} / 2 \times 10^{-19} = 2\%$. In other words the light emitted by the LED at room temperature occupies a spectral region centered at about 0.85 μm and having a spectral width of $\pm$ 2% of 0.85 μm or $\pm$ 17 nm. The nominal optical frequency is about 3×10^{14} Hz. This spectral width of $\pm$ 2% corresponds to $\pm 6 \times 10^{12}$ Hz (i.e., a bandwidth of around 12 THz). Thus it is clear that the band of frequencies emitted by the LED is much larger than the bandwidth of an information-bearing signal which might modulate the power emitted by the LED.

When we apply a forward bias current to the LED holes and electrons are injected into the junction and combine to produce photons of light. Figure 3-4 illustrates the characteristic of total output power vs applied forward bias current for a typical high radiance (efficient light emitter) GaAlAs LED. We see that for moderate applied currents (below

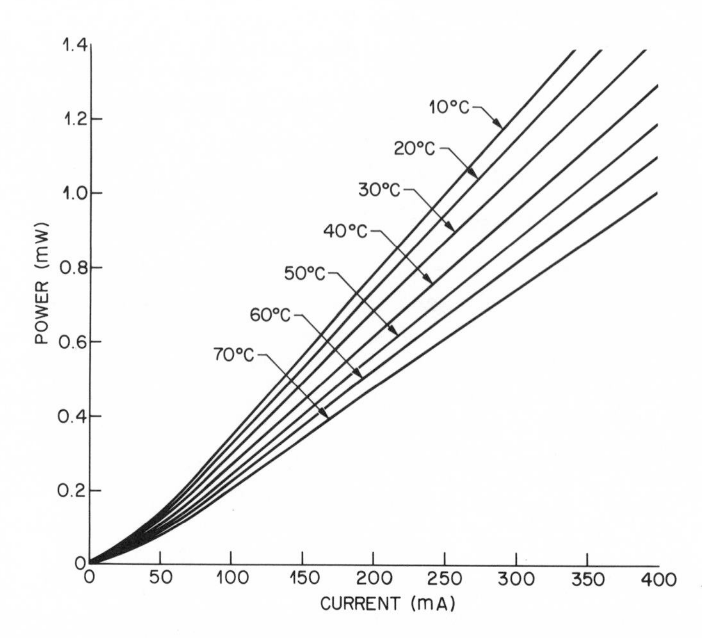

Figure 3-4 Total Output Power vs Applied Current for a Typical GaAlAs LED.

a few hundred milliamperes in this case) the output light power increases approximately linearly with the applied current. Above this level of drive, heating effects cause saturation of the light output vs current drive characteristic. We also observe that the slope of the curve decreases when the device ambient temperature increases. Higher device temperatures reduce the efficiency of conversion of hole–electron pairs to photons of light (at higher temperatures more hole–electron pairs combine nonradiatively to produce heat). Figure 3-5 shows a similar characteristic for a long wavelength (1.3 μm) InGaAsP LED. Note that the temperature sensitivity is stronger for this material.

In addition to the dc input–output characteristic of the device, we are also interested in the modulation capabilities. Modulation is accomplished by varying the drive current. The varying drive current causes the emitted output power to vary in response. If we vary the drive current slowly the output light power will follow reasonably faithfully (there is some non-linearity in the input–output characteristic, but we can neglect it in this discussion). However, if we vary the drive current rapidly the output light power may not be able to track these variations. This frequency

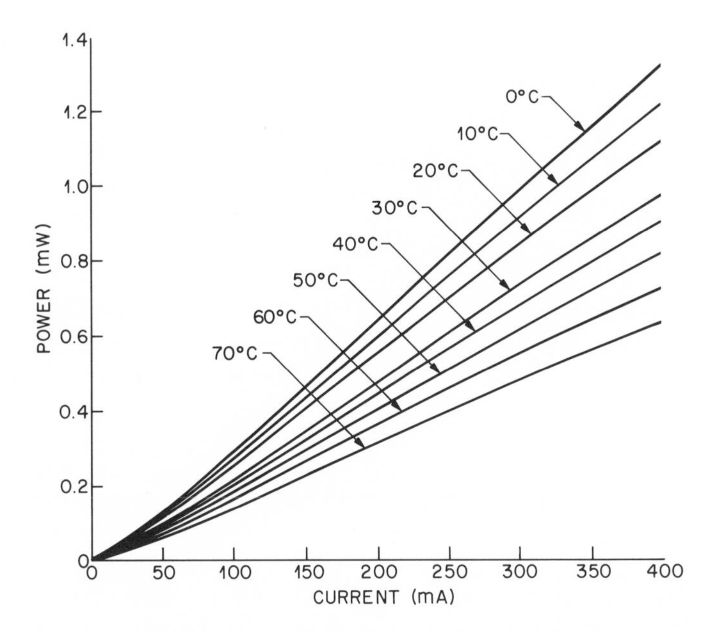

Figure 3-5 Total Output Power vs Applied Current for a Typical InGaAsP LED.

response limitation can be caused by ordinary circuit limitations (capacitances and inductances which prevent the high frequencies in the modulated current applied to the device terminal conductors from reaching the junction as injected current variations) and by the intrinsic time constants of the device itself. Circuit limitations can be minimized by careful transmitter design (see Section 3.3), minimization of lead lengths, device capacitance, and series resistance, etc. The intrinsic device modulation speed limitation is associated with the recombination lifetime of a hole—electron pair in the junction.

If a given number of hole—electron pairs are injected into the junction at some time T_0 then this number of pairs will decay exponentially with a decay time constant known as the recombination lifetime T_r. The recombination lifetime of a material can be decreased by adding dopants, which usually (but not always) increase the rate of nonradiative recombination. This reduces the conversion efficiency (ratio of light output to electrical power input) of the device. Recently results have been reported of doping techniques which can reduce the recombination lifetime while maintaining a high ratio of radiative to nonradiative recombination. The recombination lifetime relates to the modulation bandwidth of the device through the following simple equation:

$$p(t) = C \int_{-\infty}^{t} i(\tau)\exp\left[(\tau - t)/T_r\right] d\tau$$

$$B_{\text{modulation}} = 1/(2\pi T_r) \tag{3-1}$$

where $i(t)$ is the applied current, $p(t)$ is the modulated emitted light, and C is a constant. We see that the emitted light is a low-pass filtered replica of the drive current with a 3-dB (0.707 amplitude) rollof at frequency $1/(2\pi T_r)$ Hz. Note that the current, $i(t)$, in (3-1) is assumed to be positive for all times t.

The light emitted by the LED can be most simply characterized in terms of its total power, but more details are needed to understand some of the system limitations which will be described in Part 2 of the book and in Chapter 6 below. We have already mentioned the fact that the device emits light over a band of wavelengths. The device also emits light simultaneously and independently in a large number (typically thousands) of field patterns. In a sense, the LED acts like a large number of independent light emitters all operating in the same light-emitting cross-sectional area. Consider a light-emitting diode with a circular light-emitting area of diameter D operating at nominal wavelength λ. Within this area one can fit N complex field patterns $E(x,y)$, where x and y correspond to the cross-sectional coordinates, which are orthogonal. That is they satisfy the following equations:

$$\int E_i\ (x,y)\ E_j^*\ (x,y)\ dx\ dy = \begin{cases} 1 & \text{if } i = j \\ 0 & \text{if } i \neq j \end{cases} \tag{3-2}$$

Each of these field patterns corresponds to light emitted in a different direction relative to the direction perpendicular to the emitting area. In a sense, these field patterns are very much like the modes guided in a multimode fiber. The light emitted in each of these LED spatial modes is in a random phase relationship to the light emitted in the other modes. Furthermore the light in any mode has a random amplitude and phase which changes every T_s seconds, where T_s is the reciprocal of the LED spectral width. The fact that the emitted light in each mode randomizes its amplitude and phase every T_s seconds, and the emitted spatial modes have statistically independent amplitudes and phases is the reason we have referred to the light emitted by this device as noise. We shall see in Chapter 6 what some of the implications of these noiselike properties are.

When an LED is modulated, there is a small delay between the modulation waveform applied to the drive current and the corresponding response in the output power vs time waveform. It has been observed that if one examines narrow bands of wavelengths within the LED output spectrum with an optical spectrum analyzer (monochromator) one finds that this delay is not uniform across the LED output spectrum. This phenomenon is called chirping.

The curve of optical output power vs electrical drive current of the LED is slightly nonlinear (below saturation). It has been observed that if one uses an optical system to select portions of the total output field emitted by the device then there is a lack of tracking of this nonlinearity between the selected portions. That is, the nonlinearity in the input drive current vs optical power produced characteristic may not be the same for different spatial modes or combinations of spatial modes emitted by the LED.

3.1.1.2 Input–Output Characteristics of Lasers

When current is applied to a semiconductor injection laser, it behaves at first like a light-emitting diode. However, when the current reaches the threshold value, the process of stimulated emission begins to dominate the LED process of spontaneous emission, and the device begins to oscillate (laser action begins). Figure 3-6 shows a typical set of curves of light output vs applied current for a GaAlAs injection laser. We can see the temperature dependence of the threshold current and some slight nonlinearity on the light output vs applied current characteristic above threshold.

For most modern injection lasers used for optical fiber communication systems the light emitted by the device is in a well-defined spatial field pattern. In the early days of injection lasers (early 1970s) the device designs were such that this was not necessarily the case. A laser which can emit light in more than one spatial field pattern will typically show a kink (sharp bend) in its curve of output power vs drive current, corresponding to the onset of oscillation of a second spatial mode (field pattern). If a device can oscillate in several spatial field patterns (transverse modes) it will tend to divide its total output power in an unpredictable manner (randomly) amongst the allowed modes, and this subdivision will tend to change randomly in time.

The laser will oscillate at a frequency which corresponds to a resonance of the optical cavity. Refer to Figure 3-7. The wavelength of the light in the cavity is the free space wavelength divided by the index of refraction of the cavity material (e.g., 3.5 for GaAs). The condition for resonance is that the round trip length of the cavity, $2L$, be a multiple of the wavelength of the light in the cavity at the frequency of oscillation.

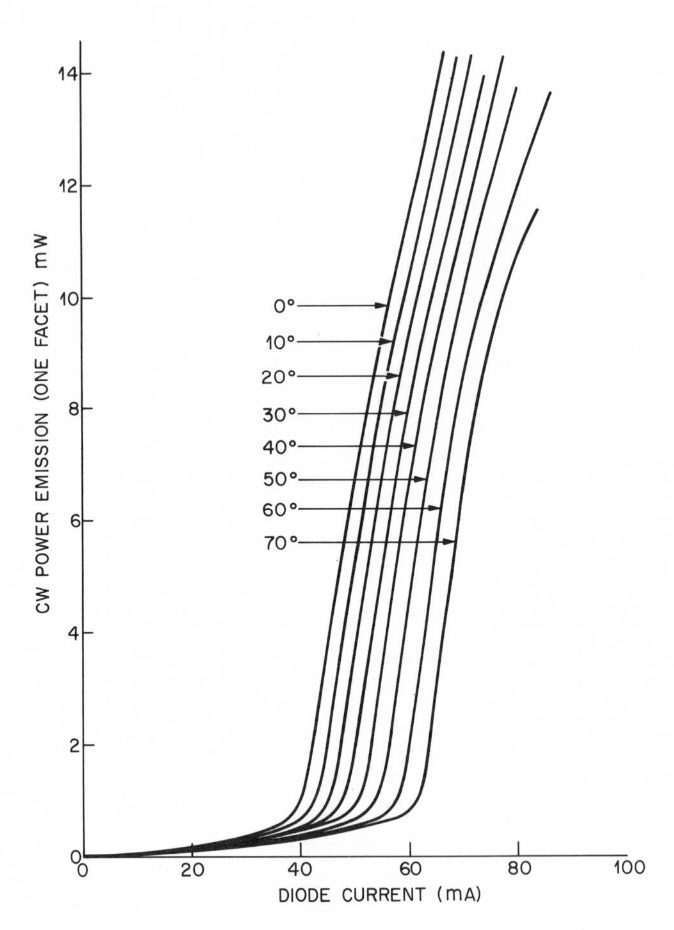

Figure 3-6 Total Output Power vs Applied Current for a Typical GaAlAs Injection Laser.

Consider a cavity of length 100 μm made of GaAs and oscillating at a nominal free space wavelength of 0.85 μm. The wavelength in the cavity is 0.28 μm. Thus the round trip length of the cavity is 200/0.28 = 706 wavelengths. The cavity will also resonate at a free space wavelength where the round trip length of the cavity is 706 $\pm$ 1 wavelengths, corresponding to 0.85 μm $\pm$ 1.2 nm. Indeed, the cavity will have a series of resonant wavelengths spaced 1.2 nm apart. If the cavity is 200 μm long, then the spacing between cavity resonances will be 0.6 nm. What mechanism will determine the particular wavelength of oscillation? In today's lasers there is a weak selection mechanism corresponding to the curve of gain in the cavity vs wavelength associated with the active layer. However, this gain vs wavelength curve has a full width to half maximum of about 25−50 nm (corresponding to the width of the spontaneous emission). Thus the round trip gain afforded to two wavelengths near the peak of this curve, but spaced by only a few nanometers, is very nearly the same. As a result of this, typical lasers can oscillate at a number of cavity resonances. The frequency of oscillation can jump from one resonance to another in an unpredicatable fashion, and the device can oscillate in several resonant longitudinal modes (several wavelengths) simultaneously, with its output power divided unpredicatably between these modes. The impact of this random multifrequency operation will be discussed in Sections 6.1 and 6.2 below. There are a number of methods which have been proposed to stabilize the frequency of operation of an injection laser. These approaches all provide for a frequency selection mechanism of sufficiently high selectivity to resolve the subnanometer spacing between normal cavity resonances. The methods include cleaving the laser cavity into two coupled cavities having slightly different sets of resonant modes, adding an external mirror to form a second cavity outside of the laser, and using corrugations formed on the laser surface to form a frequency selective mirror. In the multicavity approaches the laser will oscillate at a frequency which is common to all (both) cavities. The multicavity approaches tend to force the

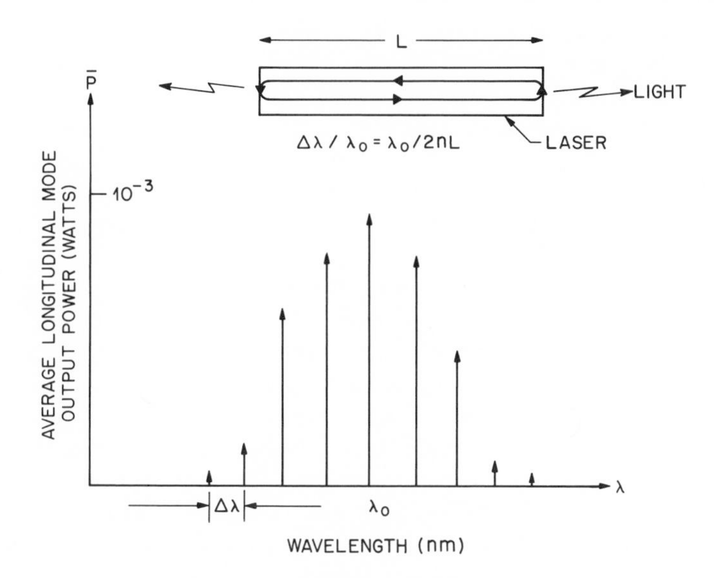

Figure 3-7 Laser Longitudinal Mode Spectrum.

laser into a single frequency of oscillation, but stabilizing that frequency of oscillation (preventing occasional jumps), in the presence of modulation, temperature changes, and bias current changes, requires complex driver circuitry. Fabrication of the distributed mirrors on the surface of the device presents difficult challenges, but much progress has been reported recently.

In addition to multifrequency oscillation, lasers are also observed to suffer noiselike fluctuations in their total output power and in the power in any particular longitudinal mode (frequency). Recently there has been an attempt to understand this behavior more clearly by studying the equations which describe the relationship between carrier (holes and electrons) populations and photon populations within the laser cavity. Attempts have been made to model the laser as a near-unity-gain amplifier of spontaneous emission, again using the above-mentioned equations. Simulations based on these models have predicted some of the observed noiselike behavior of laser emission. These noiselike fluctuations limit the signal-to-noise ratios which can be achieved at the outputs of communication links which employ lasers. This will be discussed further in Chapter 6 below.

Modulation of the laser output power is accomplished in most present applications by varying the drive current. (External modulation techniques will be discussed in Section 5.2.) Typically lasers are operated with a fixed amount of drive current to bring the device just below the threshold of lasing. Incremental current is then added to modulate the device from low output (off) to high output (on) for digital applications. For analog modulation, the device is operated with a bias current which places it at an intermediate level of output power. The device is then modulated above and below this level of output by a superimposed modulating current.

In digital applications the injection of a fixed bias current serves two purposes. First, there is a delay of several nanoseconds to bring a typical laser from zero applied current to the threshold of lasing. By applying a fixed bias current, slightly below the lasing threshold current, this delay can be avoided. The response speed of the device to the incremental currents that are added to the fixed bias can be a small fraction of a nanosecond. Thus modulation rates of several gigahertz are possible, depending upon the details of the device. Second, it is easier to produce high speed digitally modulated currents at a level equal to the incremental values, rather than to modulate the current from zero.

As in the case of the LED, the modulation limits are imposed not only by the intrinsic time constants of the interactions between carriers and photons but by circuit limitations as well. High speed modulation of lasers requires circuit design techniques which account for parasitic inductances, device capacitance, and series resistance.

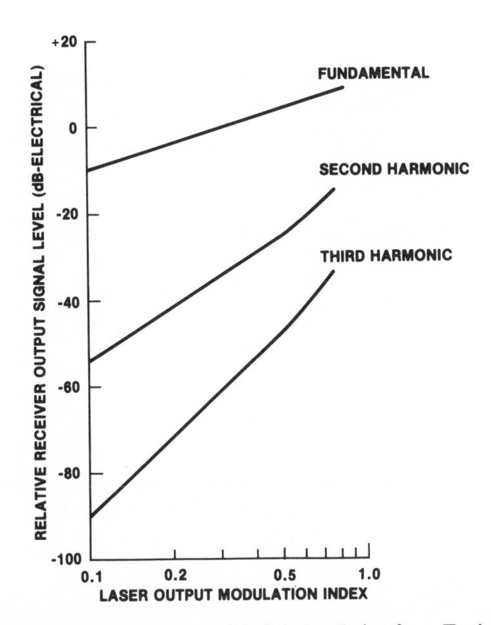

Figure 3-8 Laser Response Harmonics vs Modulation Index for a Typical GaAs Laser.

Typical lasers have a current input—light output characteristic which appears to be linear above threshold. However, if one modulates the current sinusoidally about an operating point above threshold one observes harmonics in the detected output power. Figure 3-8 shows some characteristics of a typical injection laser, demonstrating how the amplitudes of the second and third harmonics increase with the modulation index (amplitude of modulation relative to the level above threshold). These curves can be very sensitive to the operating point. For example if the device is operated near a point of inflection, then the second harmonic may be very small, for small modulation indices. The limited linearity of lasers makes the design of analog modulation methods difficult.

3.1.2 Fabrication of LEDs and Lasers

Both LEDs and lasers are fabricated by batch processes where a number of device chips are formed on a wafer using photolithographic techniques. Layers of material are deposited to form the heterojunctions and p-n junctions starting with a wafer of GaAs or InP. The growth processes used are liquid phase epitaxy, vapor phase epitaxy, and molecular beam epitaxy. In liquid phase epitaxy the growing wafer is brought in contact with a series of supersaturated liquids which cause new crystalline

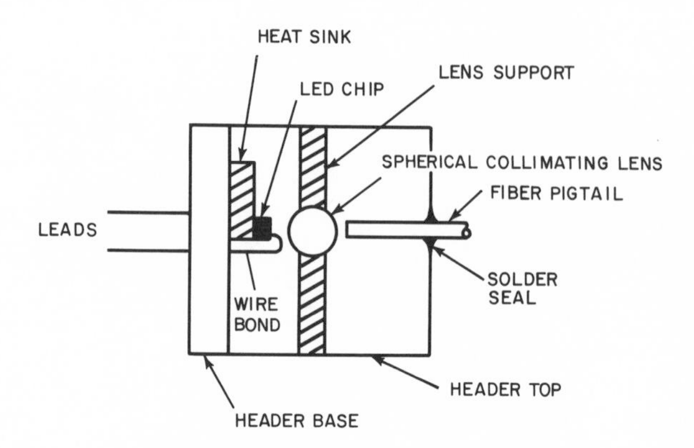

Figure 3-9 Schematic of a Packaged LED.

Figure 3-10 Photograph of an Injection Laser on its Heat Sink. The Laser is on a Heat Sink Directly Below the Bond Wire at Center. The Object to the Right of the Heat Sink is a Grain of Salt. Reprinted by Permission of Bell Communications Research and AT&T Bell Laboratories. Circa 1975.

Figure 3-11 Photograph of an Injection Laser Emitting Light with a Sewing Needle in the Background. Courtesy of RCA Laboratories.

material to grow on the surface. Precise control of the composition and temperature of the melts is required. In vapor phase epitaxy the wafer is exposed to gas mixtures of varying composition, which form new crystalline layers on its surface. In molecular beam epitaxy, as the r ne implies, the wafer is exposed to beams of molecules, evaporated from heated molecular sources, which can be turned on and off.

In the process of fabrication of the wafer containing individual devices various etching procedures may be used, either after deposition or as an intermediate step, to implement the details of the device geometry. After the wafer is fabricated, individual chips are diced from it. These must be bonded on heat sinks, bonded to conducting wires, and aligned with their output fibers. At present, this packaging operation is a substantial portion of the cost of fabricating useful sources. Figure 3-9 shows a schematic of a packaged LED. Figure 3-10 shows a picture of a laser chip on its heat sink. Figure 3-11 shows a photograph of a laser emitting light next to an ordinary sewing needle.

The fabrication of a reliable device requires not only careful control of the wafer growth process, but attention to many details in the bonding (to minimize strain) and fiber alignment procedures, as well as proper sealing of the finished package against contamination. The difficulty in implementing all of these details in a production environment may explain some of the discrepancies in the reliability of field devices compared with laboratory results. These discrepancies tend to diminish with production/field experience, which leads to improvements in manufacturing controls and methods.

3.2 Coupling Sources to Fibers

In order for the output power of an optical source to be useful in a fiber optic system application some fraction of it must be successfully and stably coupled to a fiber.

3.2.1 Coupling LEDs to Fibers

The output of a typical LED is generally emitted from a surface whose area is equal to or larger than that of a multimode fiber (diameter equal to 50 μm or larger), and generally fills a solid angle much larger than the acceptance solid angle of a multimode fiber (the range of angles captured in the core via total internal reflection). Figure 3-12 illustrates this typical situation in two dimensions. Note that with the fiber butted up against the emitting area a substantial fraction of the emitted rays are not captured because their angles relative to the fiber axis are too large, or because they do not strike the core (using the geometric optics model as

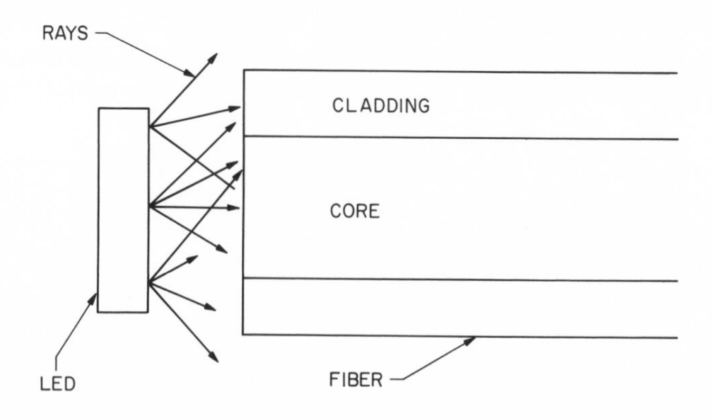

Figure 3-12 Coupling an LED to a Fiber Without a Lens.

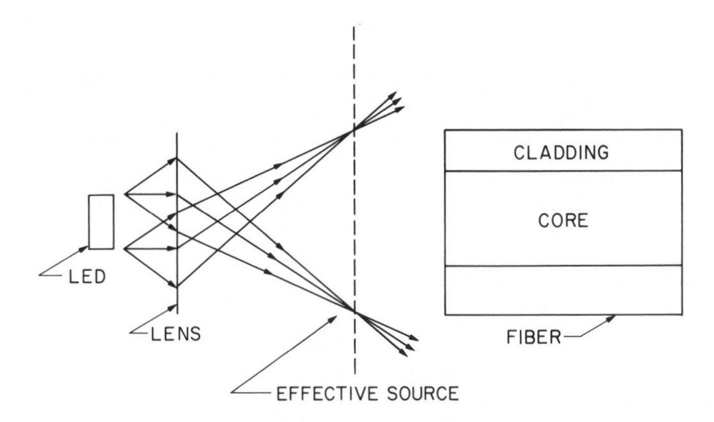

Figure 3-13 Coupling an LED to a Fiber Using a Lens.

shown). At first one might assume that the fraction of coupled light could be improved by the use of a lens. However, this is generally not the case. The LED is emitting independently in thousands of spatial modes (orthogonal field patterns), while the multimode fiber can guide only a few hundred spatial modes. One can show that the coupling efficiency from the LED to the fiber can, at most, equal the ratio of the number of guided fiber modes to the number of modes emitted by the LED. It is possible to have a much lower coupling efficiency, and sometimes lenses are necessary to approach this theoretical maximum. The situation illustrated in Figure 3-13 gives an example of how a lens can be used to improve the coupling efficiency. The light-emitting area, in this example, is smaller than the fiber core. The lens images the emitting area into an "effective source" whose area is larger but whose range of angles is smaller. The lens trades off unfilled area against noncaptured high angle emissions.

A lens can also be useful when it is not practical to bring the fiber close to the light-emitting area. In this case the lens images the light to a position away from the emitting surface. However, the constraint on coupling efficiency (known as the law of brightness) cannot be violated. In effect, the source emits a certain amount of power per spatial mode. The total power that can be coupled into the fiber is the power per spatial mode multiplied by the number of fiber modes.

The constraint on coupling efficiency between LEDs and fibers does have a positive side. If, on the one hand, the coupling efficiency is only 2%, there are, on the other hand, roughly 50 independent positions which achieve this coupling efficiency. Thus alignment is not as critical as it is in the laser-fiber case to be described below.

3.2.2 Coupling Lasers to Fibers

Most modern injection lasers emit light in a well-defined spatial field pattern which remains stable in time, with modulation, aging, etc. This implies, in principle, that the laser can be coupled with 100% efficiency to any fiber— even a single mode fiber. However, the field pattern emitted by the laser is not matched to the field pattern that is accepted by a single mode fiber, or any linear combination of field patterns accepted by a multimode fiber. Thus, if the fiber end is simply butted up against the laser, the coupling will be relatively low (a few percent for a single mode fiber, and a few tens of a percent for a multimode fiber). To correct this mismatch problem a field-matching device must be placed between the fiber and the laser. For multimode fibers this can take the form of a crude lens melted on the end of the fiber. For single mode fibers a popular technique for implementing the field-matching device is to heat and taper the fiber end. For properly aligned interfaces coupling efficiencies from lasers to multimode fibers are typically over 50%. Coupling efficiencies from lasers to single mode fibers are typically 15%–20%, with somewhat higher results reported in the laboratory.

There are noise problems caused by reflections from the coupled fiber back into the laser. These problems can be particularly troublesome in sin-

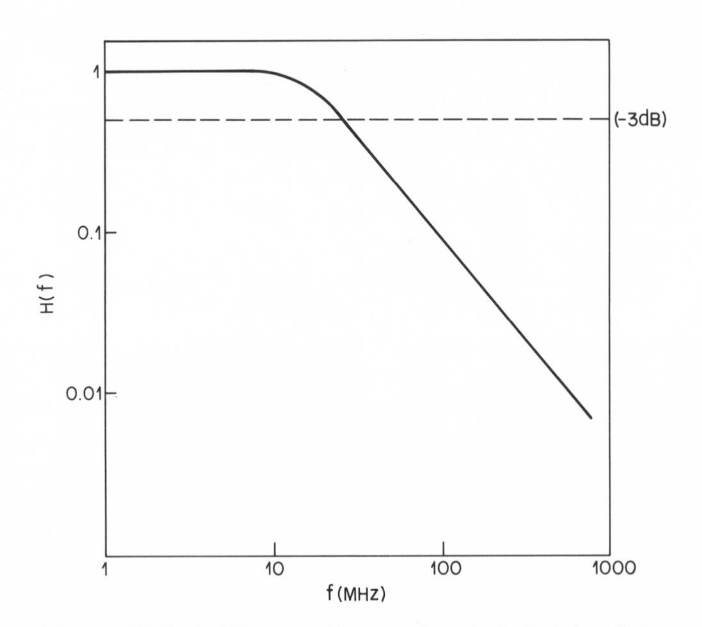

Figure 3-14 Typical Frequency Response for a Light-Emitting Diode.

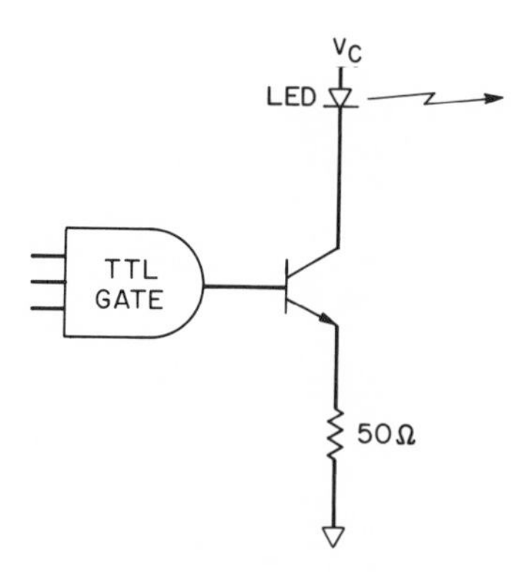

Figure 3-15 Simple LED Driver Circuit.

gle mode fiber systems, which typically operate at higher modulation speeds (and are, therefore, more susceptible to rapid fluctuations). In designing the interface between the fiber and the laser some compromise may be made to minimize reflections at the expense of coupling efficiency.

The alignment of the laser and the fiber is critical and must be stable with temperature (of the laser package) and time. It is possible to design the interface to sacrifice some coupling efficiency in order to reduce sensitivity to mechanical alignment errors.

3.3 Transmitter Modules

In order for an LED or a laser to be useful in a system application it must be successfully interfaced to the electronic terminal which is generating the modulating signal. Important considerations in the design of a transmitter module which includes the electronic interface circuitry are power consumption, modulation bandwidth, available power supply voltages, transient and overvoltage protection, impedance matching, and stabilization against temperature and aging effects. For analog modulation applications one is also concerned with flatness of frequency response and linearity.

3.3.1 Driver Circuitry for LEDs

Figure 3-4 shows a typical dc characteristic of light output vs. drive current for a light-emitting diode. Figure 3-14 shows a typical small signal frequency response of a light-emitting diode (the modulation depth of the detected output power normalized by the modulation depth of the applied current). A very simple driver interface circuit for an LED, suitable for modest modulation rates, is shown in Figure 3-15. Here the single transis-

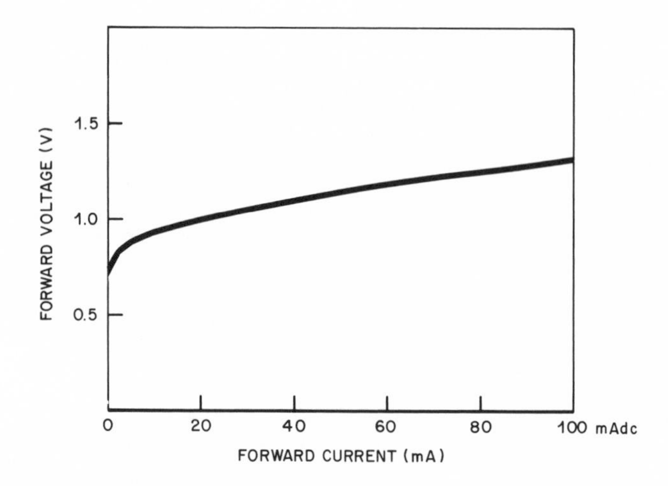

Figure 3-16 Voltage vs Applied Current for a Typical InGaAsP LED.

tor acts as a current amplifier allowing a relatively high impedance terminal to produce adequate current to modulate the LED. This circuit is suitable for analog or digital modulation. If we assume that the output of the TTL gate is 5 V in the on state, and if we assume a 0.7 V drop from the base to the emitter of the driver transistor, then we obtain 4.3 V across the 50-Ω resistor. This implies 85 mA of current through the LED for a transistor current gain of 100. Figure 3-16 shows the voltage vs applied current characteristic for a typical InGaAsP LED. We see that the 85 mA forward current results in a voltage drop across the LED of about 1.3 V. Adding up the voltages across the emitter resistor and the LED we see that

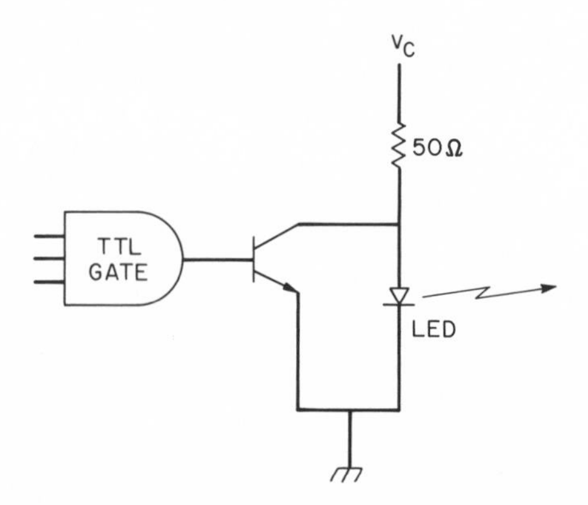

Figure 3-17 Shunt-Type LED Driver Circuit.

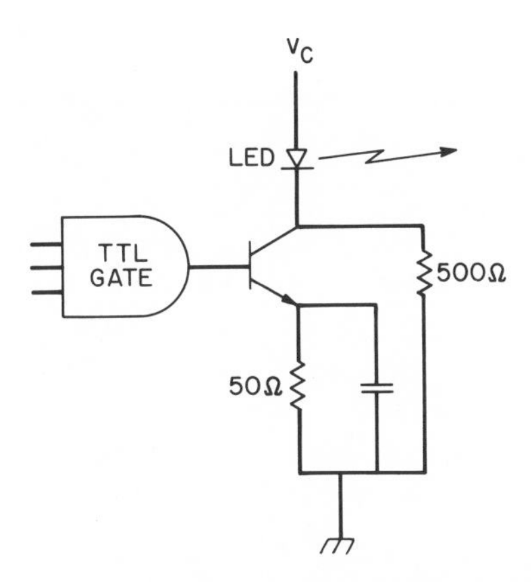

Figure 3-18 LED Driver with Speed-up Circuitry.

we require about 7 V for V+ in order for the circuit to operate as described (no transistor saturation). Figure 3-17 shows a configuration in which the driver interface transistor shunts current away from the LED. The shunt design also allows the LED "space charge" capacitance to discharge relatively rapidly through the 50-Ω resistor, reducing pattern-dependent modulation effects associated with this space charge in digital modulation applications. (In Figure 3-15 it is difficult for this capacitance to discharge rapidly when the transistor turns off, unless additional circuit components are added for this purpose.) In addition, during turn-on the LED is driven by a low impedance source which can charge its space charge capacitance more rapidly than the current source of Figure 3-15.

Figure 3-18 shows a popular approach for speeding up the turn on time of the LED by applying a larger initial current. In addition, an unmodulated component of current is drawn through the LED to keep its space charge capacitance charged near the turn-on voltage, thus reducing the effects associated with a long discharge time. In general driver designs which attempt to squeeze out more modulation bandwidth from an LED consume increasingly large amounts of power.

3.3.2 Driver Circuitry for Lasers

Driver circuitry for laser sources is typically more complex than that for LED sources for two reasons. First the temperature dependence of the laser threshold must be accommodated. Second, lasers are typically used at higher modulation speeds. (Lasers are also more prone to damage from transients.) A typical set of input drive current—light output characteristics for an injection laser is given in Figure 3-6. In practical operation it is necessary to apply an adjustable bias current to keep the laser emitting at a fixed level of short-term-average output in the presence of threshold variations, caused by temperature changes and aging.

This is often accomplished by monitoring the short-term average output power with a detector on the back face of the laser (the fiber is attached to the front face) and using this short-term average output indicator in a feedback loop to adjust the bias. Figure 3-19 shows a practical circuit for accomplishing this. Note that the short-term average output indication is compared to a reference derived from the modulating signal. In this way, if the modulating signal itself contains a changing short-term average value, this will not be tracked out by the control circuit. Furthermore, if the modulating signal calls for the laser to be off for an extended time, the control circuit will not interpret this as a result of inadequate bias. The adjustment of such a circuit for proper tracking with the reference derived from the modulating signal is a nontrivial matter.

In some laser driver designs a thermoelectric cooler is used to adjust the threshold current value of the laser, rather than adjusting the bias, to implement a controlled average output power. Cooling the laser reduces its threshold as shown in Figure 3-6.

Figure 3-20 shows a balanced driver circuit which can supply the incremental current to modulate the laser. In this circuit, designed for digital modulation, a current source is switched between the collectors of two transistors, one of which contains the laser. The bias is controlled to have a value just below the threshold current. The incremental modulation current drives the laser from this nearly off level to the "on" state. An on—off ratio of around 10:1 is typically obtained for the light coupled into the fiber, using this type of configuration. (See also Section 7.3, Figure 7-21.)

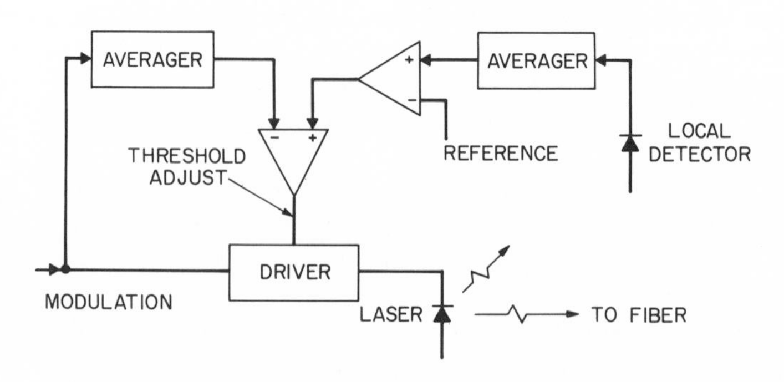

Figure 3-19 Practical Laser Driver Circuit with Feedback Bias Control.

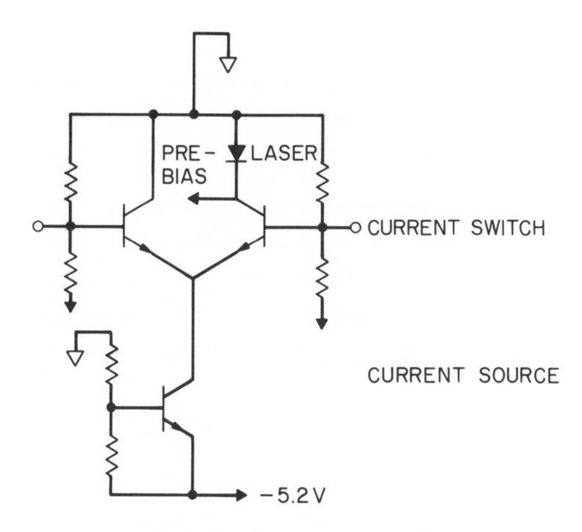

Figure 3-20 Balanced Laser Modulating Circuitry.

Recently, as interest has increased in the implementation of stabilized single frequency lasers and (more exotically) in lasers which are stabilized in frequency to an extent suitable for coherent systems (to be discussed in Section 14.2) laser driver circuitry has become potentially more complex. Essentially, provisions must be added to monitor the spectral content or the frequency of the source and to adjust the bias current, temperature, or some other control mechanism to provide the required stabilization.

Interest in very high frequency modulation of lasers (beyond 1 GHz) has led to the use of microwave circuit design techniques to properly match the laser impedance characteristics and to compensate for parasitic components in the laser and its package.

Problems

1. Using the curves shown in Figure 3-4, calculate the efficiency with which injected electrons are being converted to emitted photons at the temperature 30°C and at the current level 150 mA. Wavelength = 820 nm.

2. Using the curves shown in Figure 3-6, estimate the differential (slope) rate at which injected electrons are being converted to emitted photons in the lasing region of operation (note that the slope is not too temperature sensitive for the set of curves given). Wavelength = 820 nm.

3. A laser has a physical length (front-to-back) of 100 μm. The laser material has a refractive index of 3.5. Calculate the spacing in wavelength (nm) between resonant longitudinal modes around the nominal wavelength (free space wavelength) 1.3 μm.

4. The reflection of a plane wave at the interface between materials (fraction of power reflected) having refractive indices n_1 and n_2 is given by $(n_1 - n_2)^2/(n_1 + n_2)^2$. Calculate the reflection between the cleaved facet of a laser of index 3.5 and surrounding air of index 1. Calculate the one-way gain in dB required in order that the loss, round trip, through the laser (front-to-back, reflection from back, back-to-front, reflection from front) is 0 dB.

4

Optical Detectors and Receivers

To implement an optical communication or sensing system one requires a means of extracting information contained in the modulation of an optical signal. It is possible to perform some all-optical functions on a received optical signal, such as optical amplification. It is also possible for a received optical signal to turn on another optical signal. Examples of all-optical signal processing will be discussed in Chapter 14. However, in the optical communication and sensing systems used today (and foreseen for some time to come) information is extracted from the optical signal by converting it to an electrical signal, and by then performing electrical signal-processing functions. The device which converts the optical signal into a voltage or current is called the optical detector. The electronic circuit which interfaces the detector (with its particular characteristics and requirements) to conventional electronic terminal equipment is called a preamplifier. The combination of the detector and the preamplifier is called a receiver.

4.1 *p-i-n* and APD Detectors

4.1.1 Definition and Input–Output Characteristics

An optical detector provides a mechanism for converting optical power into an electrical voltage or current. One can imagine a variety of methods of doing this. For example, one could absorb optical energy in a device, causing it to heat up. This change in temperature could modulate a physical property (e.g., resistance) which could, in turn, be sensed by an electronic circuit. Detectors which work on absorption of optical power to produce heat (thermal detectors) are too slow in their response speeds for most optical fiber system applications (except power meters, for example).

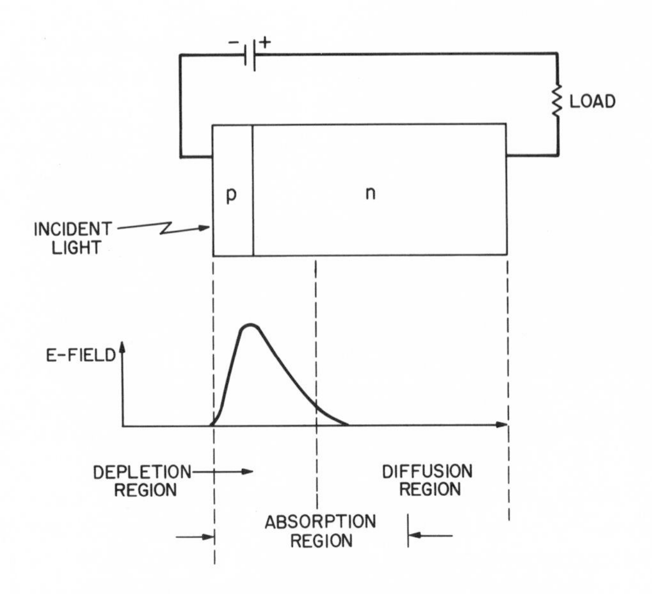

Figure 4-1 Principle of Operation of a *p-n* Diode Detector.

Another approach to implementing an optical detector is to allow the incident optical power to illuminate a semiconductor device, resulting in the generation of hole–electron pairs by absorbed photons. These pairs can, in turn, flow in the presence of an electric field to produce an observable current. This process can produce detectors with response speeds of tens of picoseconds. This is the principle of operation of *p-i-n* and avalanche photodiodes to be described below.

Figure 4-1 shows an example of a simple semiconductor device which can serve as a high speed optical detector. The device is a *p-n* diode made of a material which strongly absorbs light at the wavelength of interest. For example, for the wavelength range 0.8–0.9 μm silicon is an excellent material choice. The device is operated in a back-biased manner. In the absence of illumination only a very small current flows due to thermally generated hole–electron pairs. This small leakage current is often referred to as dark current. When the device is illuminated as shown optical energy (photons) is absorbed in the material, causing hole–electron pairs to be generated at a rate of one pair per absorbed photon. These pairs separate within the device in the presence of the local fields and produce a displacement current observable in the external electrical circuit. The total current produced in the external circuit due to the generation of a single hole–electron pair has an area (charge) equal to the charge of an electron, e. One can define three regions within the device as shown. The absorption region is the volume of space in which the light is absorbed, and extends for some depth into the device. The absorption depth depends upon the wavelength of the incident light, the material from which the device is made, and the definition of how much attenuation of the incoming light signal corresponds to absorption (e.g., 99%). The depletion region corresponds to the volume of space within the device which is depleted of mobile carriers due to the application of the reverse bias voltage (leaving behind immobile donor and acceptor ions). Within the depletion region there is an electric field which accelerates carriers. Furthermore, it can be shown that the displacement current observed in the external circuit is present only when carriers are moving in the depletion region. The volume of space which does not contain a field is called the diffusion region. Carriers within this region tend to move randomly at relatively low speeds and produce no displacement current. Thus when a hole—electron pair is generated in the diffusion region, the displacement current response will be delayed until the hole randomly finds its way to the depletion region, and is swept across the junction.

In designing the detector we want the absorption region to be long enough to absorb nearly all of the light (say 99%), but we want the resulting carriers to produce their displacement current quickly, in order to obtain a high speed device. Thus we want the depletion region to extend the full length of the absorption region. Basic semiconductor diode theory reveals that the depth of the depletion region to the right of the junction depends upon the applied voltage and upon the doping level in the n-type material. The depth increases with the square root of the applied voltage, and it increases inversely with the square root of the doping level. Practical constraints limit the voltage which can be applied to a few hundred volts. Therefore, to obtain a deep depletion region, the n-type material to the right of the junction is fabricated so as to be very lightly doped. In fact, it is so lightly doped that it is nearly an intrinsic or i-type material. It is difficult to make a good nonrectifying metallic contact to an i-type material. Therefore a heavily doped n-type material is appended to the end of the structure as shown in Figure 4-2. Thus we end up with a p-i-n (p-doped, intrinsic, n-doped) design.

The performance of a detector is measured by the efficiency with which it converts optical power to electrical current, and by its speed of response. (When one begins to design a system in detail, parameters such as dark current, junction capacitance, and linearity are also important.) The conversion efficiency is characterized by one of two equivalent measures. The first measure is the fraction of incident photons which produce electron—hole pairs. This is a number less than or equal to unity and is called the quantum efficiency of the detector η. For a p-i-n photodiode the

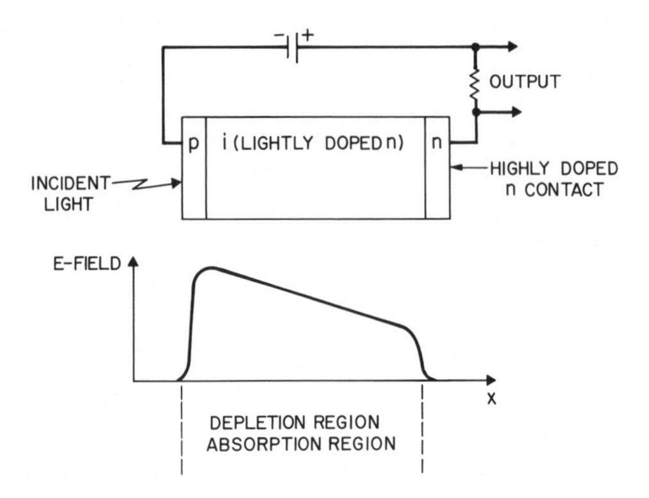

Figure 4-2 Principle of Operation of a Silicon *p-i-n* Diode Detector.

quantum efficiency can be less than unity for any of the following reasons: some incident light power (photons per second) is reflected at the interface between the surrounding medium and the detector surface; some of the light passes through the *i*-type region without being absorbed (carriers created by photon absorption in the heavily *n*-doped region recombine before producing a useful current); carriers generated in the *i*-typed region can recombine before producing a useful displacement current. For a typical detector the *i*-type region is designed to be thick enough so that light which is not reflected from the front face is absorbed in that region. Recombination in the *i*-type region is usually negligible. The second measure of conversion efficiency is the ratio of displacement current produced to incident optical power (amperes/watt). The displacement current is equal to the number of hole–electron pairs per second produced in the detector, multiplied by the electron charge, e. The incident power is the number of photons per second incident upon the detector, multiplied by the energy in a photon, hf. When units were being defined in the MKS system the electron charge turned out to be 1.6×10^{-19} coulombs (C). The energy in a photon at 1 μm wavelength turned out to be about 2×10^{-19}J. The responsivity of the detector in amperes per watt is given by $\eta \times e/hf$ where η is the quantum efficiency. The ratio e/hf has the dimensions of amperes/watt but numerically it is a number near unity (0.8 for 1 μm wavelength). Thus numerically, the quantum efficiency (dimensionless) and the responsivity (amperes/watt) are similar. For example at 1 μm wavelength, a quantum efficiency of 0.7 corresponds to a responsivity of 0.56 A/W.

The speed of response of a detector is characterized by the rise and fall times of the current waveform produced in response to an optical pulse with a fast rise and fall time when the detector is biased with a low impedance (typically 50 Ω) circuit. The response speed is limited by the time it takes for carriers to cross the depletion region (assuming the depletion region encompasses the absorption region). Typical silicon detectors designed to operate over the wavelength range 0.8–0.9 μm have a response speed of about 0.5 ns (10%–90%). If the device is optimized to work at only the shorter wavelengths in this band, then it can have a thinner *i*-type region and a faster response (silicon absorbs more strongly at the short wavelengths). However, the thinner *i*-region results in more capacitance for a given light sensitive area. It is possible to trade off quantum efficiency (responsivity) against response speed by making a thin *i*-type region and allowing the absorption region to penetrate into the heavily doped *n* material (where generated carriers recombine before producing current). Long wavelength detectors have been made from both germanium and InGaAsP materials. Properly selected compounds of InGaAsP can be very highly absorbing to light in the 1.0–1.55 μm range of wavelengths, allowing for a thin *i*-region with high quantum efficiency and response speeds below 100 ps. These thin devices can also operate with conveniently low voltages.

When the detector is integrated with a preamplifier to make a receiver one finds that the performance of the receiver is limited by the noises introduced by active and passive components in the preamplifier. We shall discuss preamplifier noise and its effects in more detail in Section 4.2 below. To combat the limitations in receiver performance caused by

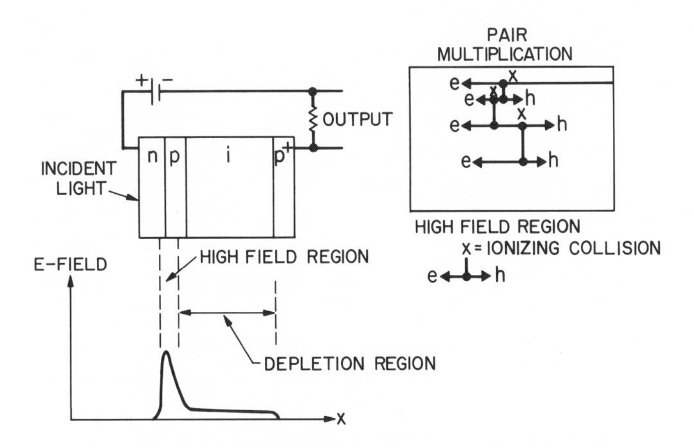

Figure 4-3 Principle of Operation of an Avalanche Photodiode.

preamplifier noise one can use a device called an avalanche photodiode (APD). The principle of the APD is shown in Figure 4-3. Doping levels are adjusted to create a region around the junction where the electric field is very high under backbiased conditions. Light which is absorbed in the *i*-type region to the right of the junction produces electron–hole pairs, the electrons of which (for silicon devices) drift into the high field region. The field levels in the high field region are sufficient to cause moving carriers to occasionally suffer ionizing collisions, where an additional hole–electron pair is produced. Carriers produced by ionizing collisions plus the original photogenerated carriers can in turn produce further pairs by the same mechanism. Thus we have a carrier multiplication process which results in a larger displacement current than we obtain with the nonmultiplying *p-i-n* photodiode. This carrier multiplication process is statistical. That is, one cannot predict exactly how many secondary pairs will result from an initial primary pair. Careful design and control of fabrication of the APD is required to produce a statistical distribution which gives the best tradeoff between average multiplication and noise associated with the statistical nature of the multiplication process. APD statistics will be discussed further in Section 4.1.3. APDs have been fabricated in silicon, germanium, and InGaAsP materials for both short and long wavelength fiber optic systems. Figure 4-4 shows a curve of multiplication factor (and equivalently responsivity) vs applied reverse bias voltage for a typical silicon APD. As reverse bias is applied, holes and electrons (mobile carriers) move away from the junction to form the depletion region. At low reverse bias, only the high field region is depleted (partially) and there are no fields in the *i*-type region to accelerate carriers produced by absorbed photons. Thus at low reverse bias the APD is a low-speed device. As the reverse bias is increased the high field region continues to deplete and the fields within that region begin to be sufficient to cause multiplication. However, at that point the device is still a low-speed detector. At larger reverse bias the high field region becomes fully depleted; and further increases in voltage deplete the *i*-type region. The device is now a high-speed detector. When the *i* region depletes, the width of the field increases, resulting in an increase in the incremental voltage required to increase the field level. Thus the curve of gain vs voltage has a kink at the point of *i*-region depletion as shown in Figure 4-4. If the reverse bias voltage is sufficiently high one obtains a condition of infinite multiplication, corresponding to sustained avalanche breakdown.

The multiplication factor of the APD changes with temperature as shown in Figure 4-4. Higher temperatures reduce the mean free path of moving carriers between nonionizing collisions, and reduce the chances of obtaining an ionizing collision. At high values of multiplication this tem-

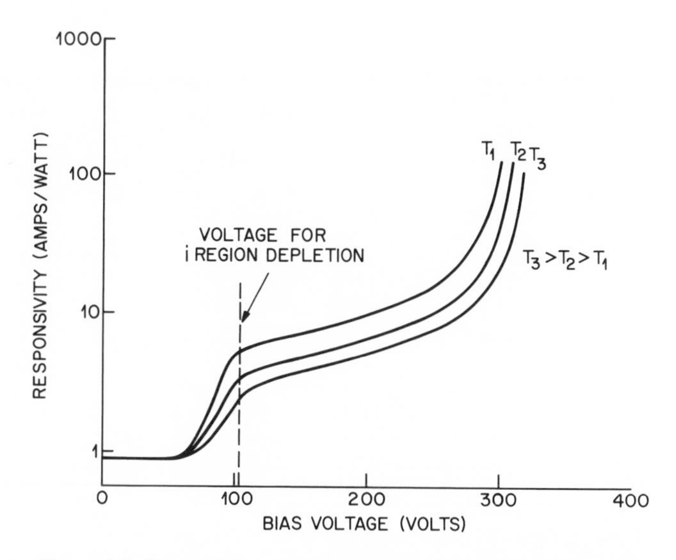

Figure 4-4 Responsivity vs Reverse Bias for a Typical Silicon APD.

perature dependence makes it difficult to bias the device for a desired level of multiplication. Receivers which incorporate APDs generally must implement some method of compensation of this temperature dependence. This will be discussed in Section 4.2 below.

Figure 4-5 shows a simplified schematic of a typical receiver incorporating a *p-i-n* or APD detector. Figure 4-6 shows the equivalent electronic circuit. The photodetection process is represented by a current

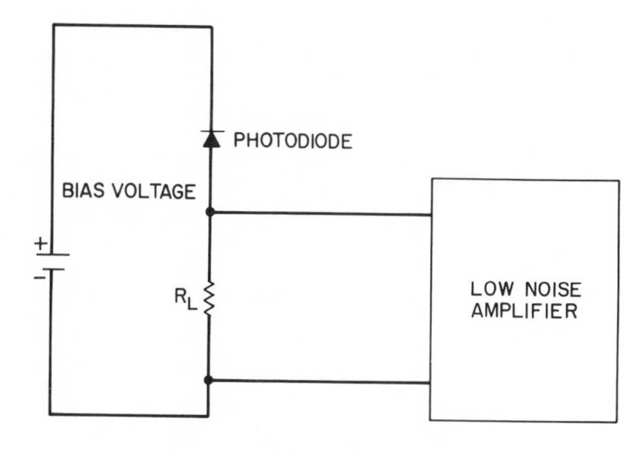

Figure 4-5 Simple Receiver Schematic.

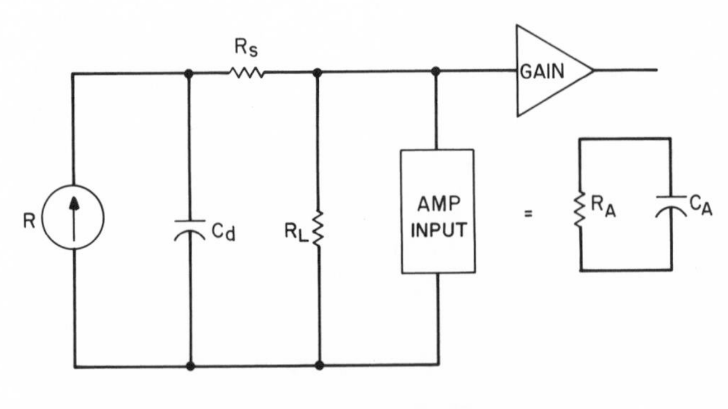

Figure 4-6 Equivalent Circuit of Simple Receiver.

source with responsivity R amperes/watt. The detector junction capacitance appears in parallel with this current source. A series resistance (typically assumed to be a few ohms or a few tens of ohms) is also shown. The preamplifier input impedance is represented by a parallel capacitance and resistance. We shall study this circuit more in Section 4.2. Note that *p-i-n* detectors are linear (R is a constant) over many orders of magnitude of incident optical power level. (They are essentially limited by noise levels at low illumination, and by power supply limitations or excessive dissipation at high optical illumination.) On the other hand APDs are somewhat

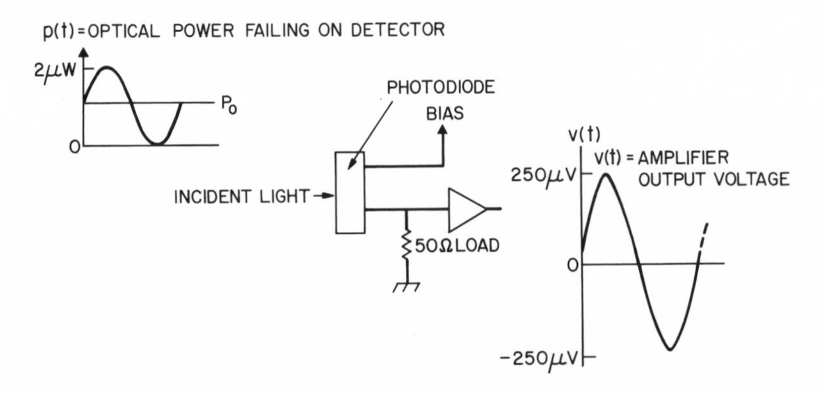

Figure 4-7 Illustration of a Photodetector in Operation.

nonlinear because the presence of carriers in the high field region can reduce the local field levels and thus reduce the multiplication.

As an example of how a photodetector might work in a circuit, Figure 4-7 provides a simple illustration. An optical signal of average power level 1 μW is shown incident upon a detector. The optical signal is modulated sinusoidally at a moderate frequency from 0 to 2 μW. The detector is in series with an amplifier having a 50-Ω input impedance and a gain of 10 (20 dB). The detector responsivity is assumed to be 0.5 A/W. The 1 μW peak sinusoidal optical modulation produces a sinusoidal detector output current of 0.5 μA (0.5 A/W). This current flows into the 50-Ω amplifier input impedance to produce an amplifier input of 25 μV peak. This results in an amplifier output of 250 μV. Further gain of a factor of 1000 (60 dB) is required to bring this up to a level of about 0.25 V.

When designing high speed and low noise receivers the capacitance of the detector is of importance. The capacitance depends upon the area of the i region, the thickness that is depleted, and the dielectric constant of the material (plus parasitic capacitances). Silicon detectors are typically fabricated with a relatively thick depletion region (100 μm or more) and a relatively large area (250—500 μm diameter). The thick depletion region is required for high quantum efficiency (to absorb the light). This thick i-region results in a low capacitance since the capacitance is proportional to the i-region area divided by its thickness. InGaAs devices are typically fabricated with a relatively thin i-region. This is possible because of the high absorption constant of this material attainable in the wavelength range 1.0—1.55 μm. This results in very fast response speed but requires a small area of the i-region (typically 100 μm diameter) to maintain a low capacitance. The capacitance of a typical unpackaged detector is below 2 pf and can be as low as a small fraction of a pf for small area devices. A small area device is naturally more difficult to couple light into.

The volume in which hole—electron pairs are generated is relatively large compared to the volumes of typical components in integrated circuits. As a result, xrays, gamma rays, and heavy particles can produce objectional amounts of dark current in environments where such radiation is contemplated. This can be particularly problematic for APDs which operate with very low photocurrents. Dark current can manifest itself in the form of shot noise (see Section 4.1.3 below) or can overload electronics in the preamplifier interfacing with the detector.

4.1.2 Fabrication of *p-i-n* and APD Detectors

p-i-n and APD detectors are typically fabricated in a batch process where many devices are produced at once on a common wafer. The layers of p, i and n type material can be formed by epitaxial growth as described

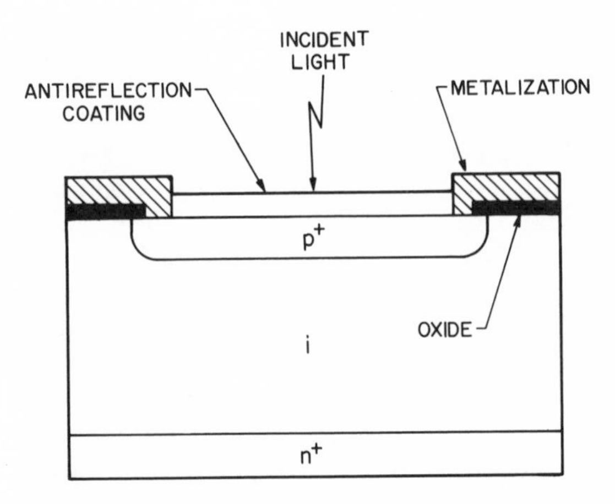

Figure 4-8 Schematic of a Typical Silicon *p*-*i*-*n* Detector.

briefly in Section 3.1.2 above, or by the diffusion of dopants into a starting substrate. Figure 4-8 shows a silicon *p*-*i*-*n* detector fabricated by diffusing dopants though a series of masks to form the *p* and *n* regions in an *i*-type substrate. An oxide layer provides insulation. Also shown is a thin layered antireflection coating applied to reduce the reflection from the surface of the device (thus increasing its quantum efficiency). Figure 4-9 shows a silicon avalanche photodiode also fabricated by diffusion processes. The high field region is formed by consecutive diffusion of *p* and *n* dopants. The artifact labeled "guard ring" is incorporated to prevent premature breakdown due to fields at the edge of the device. Uniformity and parameter

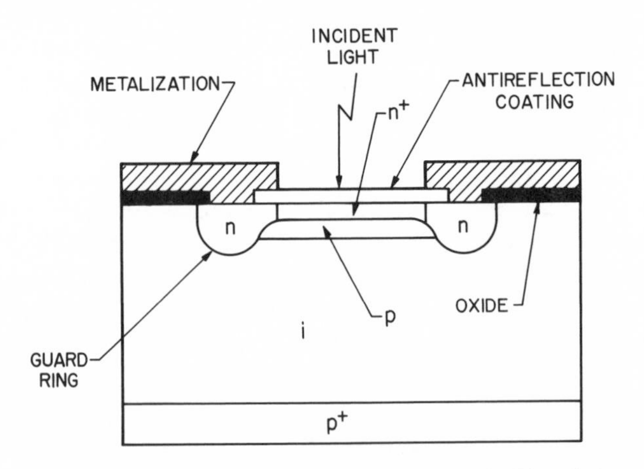

Figure 4-9 Schematic of a Typical Silicon Avalanche Photodiode.

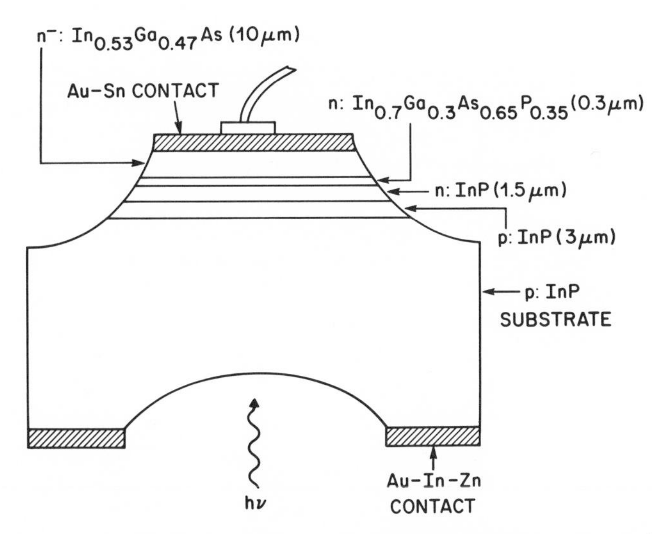

Figure 4-10 Schematic of an InGaAsP Avalanche Photodiode. Courtesy of J. C. Campbell, AT&T Bell Laboratories.

control are critical in the fabrication of useful APDs. Variations in the high field region of the local field required for breakdown (due to material imperfections) can cause excessive noise due to random occurrences of very high multiplication (microplasma breakdown) of carriers passing through those regions. Non-uniformity of the high field region multiplication characteristics adds noise by enhancing the unpredictability of the multiplication experienced by individual carriers. Poor control of the width or doping levels in the high field region can result in devices which have a slow response speed (i-region not depleted) at the desired multiplication levels, have too large a required voltage for multiplication, or which have poor multiplication statistics. Figure 4-10 shows an avalanche photodiode fabricated in the InGaAsP material system using epitaxial growth and etching processes. Here a complex layered structure is grown in order to balance a number of important factors. The i region (layer at the top labled n^-) is fabricated from a material with a narrow band-gap (energy difference between valence band and conduction band) in order to absorb the long wavelength light strongly. However the high field region (two layers below) is fabricated from a higher band gap material to prevent excessive leakage (dark) current from being generated in this volume. A transition layer between the high field region and the i-region prevents carriers from being blocked by abrupt energy level discontinuities (which form barriers).

The problem of coupling the fiber to the detector is not as difficult as in the case of the optical source. The detector has a relatively large area (compared to the fiber core) and a large acceptance angle. The fiber must be placed in reasonable alignment with the sensitive area and fixed to prevent motion from temperature induced expansion of the assembly or mechanical stress.

To obtain high sensitivity, particularly with *p-i-n* detectors, it is important to minimize the total capacitance of the detector, the input of the preamplifier, and wiring. In addition, the low level and high impedance interface between the detector and the preamplifier is susceptible to induced noise. To accommodate these constraints the detector is often mounted as a chip component on a hybrid circuit board in close proximity to the preamplifier. The entire assembly, including the attached fiber, can then be enclosed within an environmentally sealed and shielded metallic package.

4.1.3 Detection Statistics

In this section we shall review the statistics of the detection process, both for *p-i-n* and APD detectors, to understand how these statistics affect the performance of receivers.

Classical radio systems are limited by thermal background noise. Since an optical system is essentially a very high frequency radio system (fiber systems operate at around 3×10^{14} Hz) one might ask why the limiting noises in optical systems should differ from those of radio systems. The answer is a matter of approximations. Classical radio systems span the frequency range from below 100 Hz to tens of gigahertz. Over this nearly nine order of magnitude range the models used to predict the performance of radio receivers have held up pretty well. However, when we make the additional four order of magnitude leap to fiber system frequencies certain approximations begin to break down. Noise sources which were dominant at classical radio frequencies become negligible at optical frequencies (as we shall see shortly). Effects which were negligible at radio frequencies become important or dominating at optical frequencies. Consider the classical formula for thermal noise. It is given by kT watts per mode per hertz of bandwidth in a classical multimode transmission line, where k is Boltzmann's constant and T is the temperature of the medium producing the excitation of the transmission line. For example, for a low loss transmission line fed by an antenna, T would be the temperature of the medium in the field of view of the antenna. However, we can recall a result from physics, where it has been pointed out that such a flat noise spectrum implies infinite noise power if one considers the infinite possible range of frequencies. Quantum theory leads to a correction of the expression for thermal noise, and the corrected expression is given as follows:

$$N(f) = hf\,[\exp(-hf/kT) - 1]^{-1} \tag{4-1}$$

where f is the frequency, $N(f)$ is the noise spectral density per transmission line mode, and h is Planck's constant. Note that at room temperature kT is approximately 4×10^{-21} J. At 10^9 Hz hf is approximately 10^{-24} J. Thus, we see that for classical radio frequencies kT is much greater than hf. In this case equation (4-1) reduces to $N(f) = kT$; the classical expression. However, at 3×10^{14} Hz (1 μm wavelength) hf is approximately 2×10^{-19} J. In that case hf/kT is approximately 50. We see then that $N(f)$ is much less than kT at these frequencies. In fact, unless we are looking into the the sun or at some other very hot medium, thermal background noise is typically negligible in optical systems at wavelengths shorter than 2 μm, and in particular fiber systems.

If background noise is negligible, it might seem that there is no lower limit on the theoretical sensitivity of an optical receiver. However, this is not the case. A phenomenon known as quantum noise sets a lower limit on the detectable optical power level. What this lower limit is depends upon the detection problem at hand.

Suppose we have an ideal photodetector with no dark current. Suppose we illuminate the detector with a pulse of optical energy E joules. It can be shown that the number of hole–electron pairs produced within this (nonmultiplying) detector is not predictable; but follows a statistical distribution. The probability that exactly n hole–electron pairs will be produced is given by the Poisson distribution

$$p(n) = \Lambda^n e^{-\Lambda}/n! \tag{4-2}$$

where Λ is the average number of photons in the received pulse, $\Lambda = E/hf$, and e is the natural log base. From equation (4.2) we see that even though E is known in advance, we may get anywhere from zero to an infinite number of pairs produced. The average of this statistical distribution is E/hf pairs. Thus the average number of pairs produced (if we performed this experiment many times over) is equal to the average number of photons in the received pulse. The fact that the number of pairs produced is not predictable is not a defect in the detector; it is a manifestation of the phenomenon of quantum noise.

Consider the following situation. We produce, at a transmitter, an optical pulse having the shape shown in Figure 4-11 with a peak power level of 2 mW and a duration of 1 ns. The total energy in the pulse is 1 pJ (10^{-12}J). This pulse propagates through a medium (e.g., a fiber) which

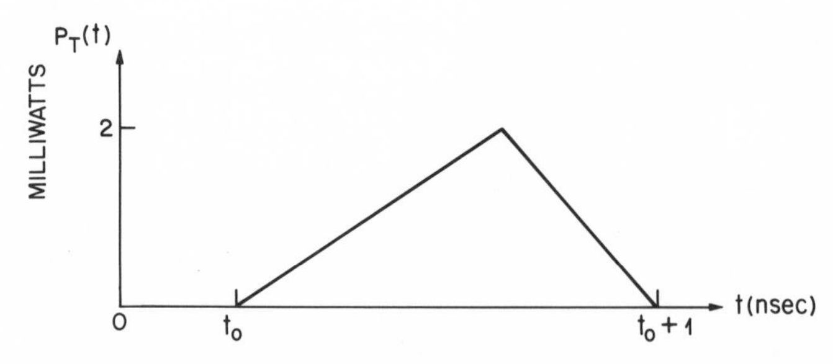

Figure 4-11 Transmitted Pulse Shape.

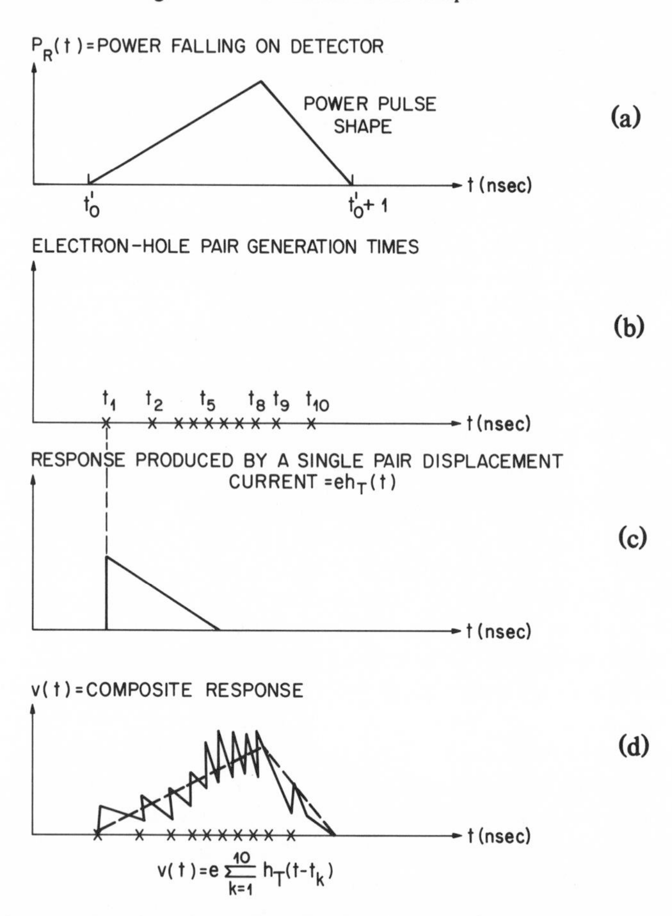

Figure 4-12 Realization of Detector Hole–Electron Pair Production Times. (a) Received Pulse Shape, (b) Times at Which Pairs are Produced, (c) Individual Hole–Electron Pair Displacement Current Response Waveform, (d) Composite Response Resulting from all Hole–Electron Pairs.

introduces a loss of 60 dB. Thus the pulse received at the other end of the medium has energy 10^{-18} J. The energy in a photon is assumed to be 10^{-19} J at the wavelength being used. Thus the received pulse contains 10 photons (in a sort of average sense). This received pulse, shown in Figure 4-12a, will illuminate a detector which will produce hole–electron pairs in response. We cannot predict exactly how many hole–electron pairs will be produced, or a what times (statistical uncertainty due to quantum effects). Figure 4-12b shows one possible realization of this experiment where 10 hole–electron pairs are produced at the times shown by the crosses. We assume that the detector produces a displacement current in response to each pair, having the shape shown in Figure 4-12c. Figure 4-12d shows the composite response from all of the displacement currents (from all ten pairs). We see that the composite response is reminiscent of the incident power waveform, but only a noisy approximation. Further, there is no arrangement of ten of the responses shown in Figure 4-12c which superimpose to produce the shape shown in Figure 4-12a.

Suppose on the other hand that the photon energy was five orders of magnitude smaller (as it would be for a microwave frequency). In that case we would have 10^6 photons in the received pulse. This would result in about 10^6 hole–electron pairs produced in the detector. With 10^6 individual responses to superimpose, we can construct a composite displacement current waveform which is an excellent approximation to the shape shown in Figure 4-12a. Thus we see that quantum effects can be important at optical frequencies and negligible at microwave frequencies and below.

Consider next the following simple communication system. A transmitter is turned on or off to produce a pulse of either E_T J or 0 J. Thus the transmitter has a perfect extinction (off–on) ratio. This pulse arrives, attenuated, at a receiver containing an ideal *p-i-n* detector and an ideal electronic preamplifier. The received pulse energy is either E_r or 0 J. The ideal *p-i-n* detector produces no dark current. Further there is no background light. Thus no hole–electron pairs are produced in the absence of a signal from the distant transmitter. The preamplifier electronics are so noise free that the displacement current of a single hole–electron pair is observable. At the receiver we use the following simple rule to determine if a pulse of light was sent from the distant end (or not). If no hole–electron pairs are produced in the detector we decide that no pulse was sent. If even a single hole–electron pair is observed (via its displacement current) we decide that a pulse was sent.

We note that if no pulse is sent by the distant transmitter, no hole–electron pairs can be produced. Therefore the only way we can make a decision error is if a pulse of optical energy is sent, but still no pairs are produced. The probability that n pairs are produced if the received energy is E_r is given by

$$p(n) = (E_r/hf)^n\, e^{-(E_r/hf)}/n! \tag{4-3}$$

Thus the probability that no pairs are produced is given by

$$p(0) = \exp[-(E_r/hf)] \tag{4-4}$$

Suppose we wish to have the probability of such an error be equal to 10^{-9}. Then by algebra we have the requirement that E_r/hf be equal to 21. That is, in order to guarantee that only one time in a billion no pairs will be produced in response to a received pulse, we require that the received pulse contains 21 photons (on the average). This requirement of 21 photons per received pulse is called the quantum limit for binary direct detection (detection with a *p-i-n* detector). If we wish to communicate a sequence of pulses at rate B pulses per second, and if on the average half the pulses are on and half are off, then the power required at the receiver is

$$P_r = 1/2 \times 21 \times hf \times B \tag{4-5}$$

For example if B is 10^8 pulses per second and hf is 2×10^{-19} J, then we require 2.1×10^{-10} W of power at our ideal receiver to produce a bit error rate of 10^{-9}.

This minimum required power is not a defect in the *p-i-n* detector, but a manifestation of the limits imposed by quantum effects. It can be shown that under very general assumptions the performance of any physically realizable receiver cannot be more than a negligible amount better than that calculated above (for binary on–off modulation).

An actual receiver will not perform as well as the ideal receiver described above. To understand why, we must examine the assumptions we made. We assumed that the transmitter was perfectly off in the "off" state. We also assumed that there was no dark current in the detector, and no background light. These assumptions guaranteed that there were no pairs produced in the detector when the transmitter was not sending a pulse. In reality, the transmitter extinction ratio will not be perfect, and there will be some dark current in the detector. However, the effects of these departures from ideal on the performance of the receiver are not large. Typically we might find that the sensitivity of the receiver would be reduced from 21 photons required per pulse to about four times that value (a 6-dB sensitivity penalty) as a result of these deviations from ideal assumptions. The assumption which is most unrealistic is that the displacement currents produced by individual hole–electron pairs produced in the

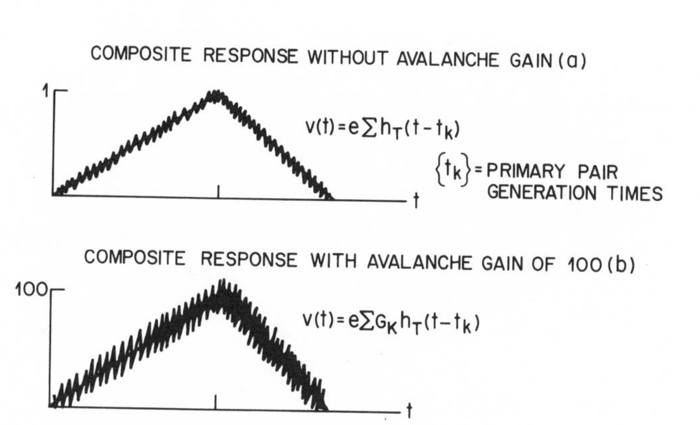

Figure 4-13 Composite Pulse Produced by a Photodiode, (a) *p-i-n* detector, (b) APD detector.

detector are observable in the background of preamplifier noise. As we shall see in Section 4.2 below, even with heroic design effort, preamplifier noise is hundreds of times as large as the response produced by a single hole–electron pair generated in the detector. Thus only the cumulative response of hundreds of hole–electron pairs is observable. For this reason the sensitivity of a typical receiver with a *p-i-n* detector is about 30 dB worse than the ideal of 21 photons required per received pulse. To obtain a performance closer to ideal, one requires a detector which produces many hole–electron pairs per detected photon.

As described above, an avalanche photodiode can produce a "bunch" of secondary pairs in response to a primary photon- generated pair. This gain mechanism can help to overcome the limitations of noises added by the preamplifier electronics. However, the number of secondary pairs produced by a given primary pair is only statistically predictable. Figure 4-13a shows the composite current response of a *p-i-n* detector illuminated by the optical pulse shown in Figure 4-11. The quantum noise effects due to the finite number of hole–electron pairs contributing to the response are visible. Figure 4-13b shows the composite response of an APD, where each primary hole–electron pair produces an average of 100 secondary pairs via collision ionization. Note that although each primary produces, on the average, 100 secondary pairs, the actual number of secondary pairs produced by any primary pair is not predictable. The waveform shown in Figure 4-13b is, on the average, 100 times as large as the waveform shown in Figure 4-13a. However, the noise on the waveform of Figure 4-13b is more than 100 times larger than that of Figure 4-13a. (The width of the noise band is larger even though the scale factor has been increased by

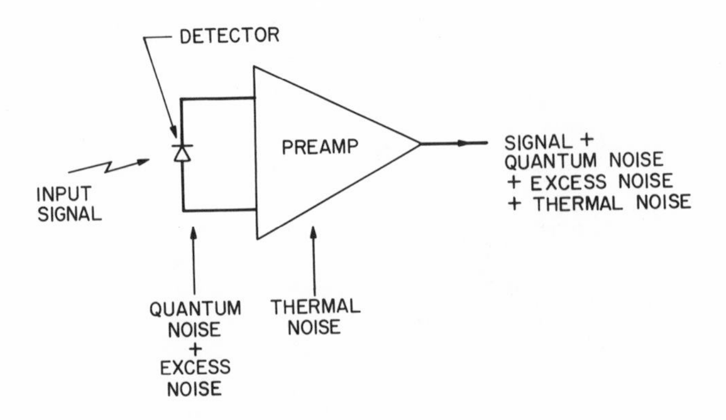

Figure 4-14 Noise Sources at the Preamplifier Output.

100.) Thus the signal-to-noise ratio is smaller in Figure 4-13b. Why then would we use avalanche gain? The reason is that in both Figure 4-13a and Figure 4-13b we have not shown any additive amplifier noise. Suppose the amplifier noise had an rms value of 5 units height on the scales used in these figures. In Figure 4-13a with the *p-i-n* detector in use, this noise would completely mask the signal (which is only 1 unit in height). In Figure 4-13b the signal is 100 units in height, and would be clearly visible in the preamplifier noise of 5 units height. (Remember, the avalanche gain takes place in the detector, and does not amplify the preamplifier electronic noise.) Thus, the net signal-to-noise ratio, taking into account the quantum noise, the enhancement of quantum noise caused by the randomness of the avalanche gain, and the preamplifier noise, is improved by the use of avalanche gain.

As the avalanche gain is increased from unity (no gain) the signal-to-noise ratio improves as described above. However, as the avalanche gain is increased to very large values one can reach a point of diminishing returns where the increased noise from the randomness of the gain mechanism may be worse than the preamplifier noise. This can be seen mathematically as follows. If we look at the output of the preamplifier shown in Figure 4-14 we see three components: the desired signal from the detector, quantum noise which is enhanced by the random avalanche mechanism, and preamplifier noise. The desired signal has an amplitude proportional to the average avalanche gain, G; and also proportional to the received optical power level P_r (watts). The preamplifier noise has some variance N_a. The enhanced quantum noise can be shown to have a variance given by the product of the received optical power level P_r, the square of the average avalanche gain, G^2, and an excess noise factor $F(G)$ which is larger than unity and grows with increasing average avalanche gain. Thus the signal-to-noise ratio, SNR, (ratio of desired signal squared-to noise variance) at the preamplifier output is given by

$$\mathrm{SNR} = aP_r^2\, G^2\, [N_a + bP_r\, G^2\, F(G)]^{-1} \tag{4-6}$$

where a and b are constants.

We see that as G is increased from unity, the signal-to-noise ratio increases until the first term in the denominator is smaller than the second term. Beyond this, further increases in G can decrease the signal-to-noise ratio because $F(G)$ increases with increasing G.

The statistics of avalanche multiplication are complex. Details are available in the references. However, expressions have been derived for the excess noise factor $F(G)$. In an APD, both holes and electrons can produce ionizing collisions. For a given distance of travel in the high field region one carrier is usually more likely to suffer an ionizing collision (is more highly ionizing). It can be shown that the best multiplication statistics are obtained if the more highly ionizing carrier initiates the multiplication process. That is, the device should be designed such that when photons are absorbed in the i-region, the more ionizing carrier drifts into the high field region. It can also be shown that the best avalanche statistics are obtained (minimum gain variance for a given average gain) when the ratio of ionizing ability of the two carrier types is large. For example in an ideal APD only one type of carrier (electrons or holes) would cause ionizations. The ratio of the probability of ionization per unit distance of motion in the high field region for holes and electrons is called the ionization ratio. If electrons are the more ionizing carrier, then we want this ratio to be as small as possible. The APD designer has some control over this ratio because the ionization probabilities are dependent upon the field levels in the high field region, and they scale differently for the two carriers. The ratio is called k (not to be confused with Boltzmann's constant) in the literature. Assuming that the multiplication process is initiated by the more ionizing carrier, we obtain the following formula for $F(G)$:

$$F(G) = \left(2 - \frac{1}{G}\right)(1-k) + kG \tag{4-7}$$

Figure 4-15 shows a plot of $F(G)$ vs G for various values of k. The performance of a receiver using an avalanche detector depends upon the details of the avalanche gain distribution. However, certain approximate expressions for the performance depend only on the variance of the avalanche gain, which is given by $G^2\, F(G)$. We will not reproduce the

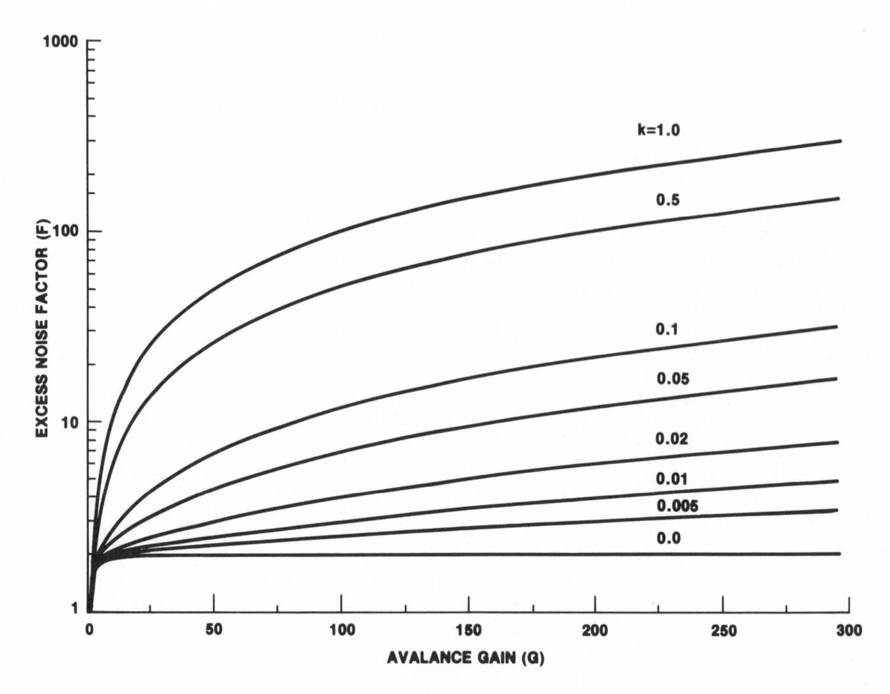

Figure 4-15 Excess Noise Factors vs Gain of APD Detectors.

complicated derivations of the avalanche gain distribution here. However, we can state a few interesting results. For an ideal unilateral gain detector, where only one carrier suffers ionizing collisions ($k = 0$ or k = infinity), the gain distribution is geometric. That is, the probability of exactly g secondaries produced by a primary pair (including the primary pair itself) is

$$P_G(g) = \frac{1}{G-1}\left[1-\frac{1}{G}\right]^g \quad \text{for } g \geqslant 1 \tag{4-8}$$

where G is the average gain (multiplication)

This is approximately an exponential distribution when G is larger than 10. When k is not 0 or infinity the distribution of the gain g is more complex, but always has an exponentially shaped tail. An approximate expression for this distribution, valid for G greater than 10 is

$$P_G(g) = \frac{(2\pi)^{-1/2}\, G^{1/2}\, F(G)\,(1/g)^{3/2}}{(F(G)-1)^{3/4}} \exp\{-g/2G[F(G) - 1]^{1/2}\} \tag{4-9}$$

where G is the average value of the gain g and $F(G)$ is the excess noise factor, given in equation (4-7). In Section 4.2.3 we shall give some results concerning the improvement in receiver performance which is obtained using APDs (vs. p-i-n detectors).

4.2 Receiver Modules

A receiver consists of the combination of a detector and a preamplifier which accommodates the characteristics and requirements of the detector to optimally interface it to conventional electronics. The objective is to build a receiver subsystem which provides a combination of high sensitivity (low required incident optical power to achieve a desired performance), adequate dynamic range (ability to accommodate a range of incident optical power levels), adequate bandwidth, and a proper interface to the remaining electronics beyond the receiver.

4.2.1 Preamplifier Design [8]

The preamplifier in an optical receiver must be designed to act as an interface between the detector and conventional electronics. In particular it must amplify the weak current produced by a photodetector; provide a low impedance level interface to following electronic stages; and it must do so with as little additive noise as possible. Preamplifiers are required in conventional microwave radio and metallic cable systems as well. However, the characteristics of the optical detector are different from those of an antenna feed or a coaxial cable. Figure 4-16 shows a simplified schematic of a resistive signal source, characteristic of many (but not all) classical systems. The signal is represented by a current source in parallel with a resistance. Also included is a noise current source associated with the thermal noise of the resistance. Figure 4-17 shows a schematic of the equivalent circuit of a photodiode. The signal is represented by a current source in parallel with a capacitance. (The series resistance of the photodiode is neglected here). If we neglect the shot noise of leakage current, there is no intrinsic noise source apparent in this model.

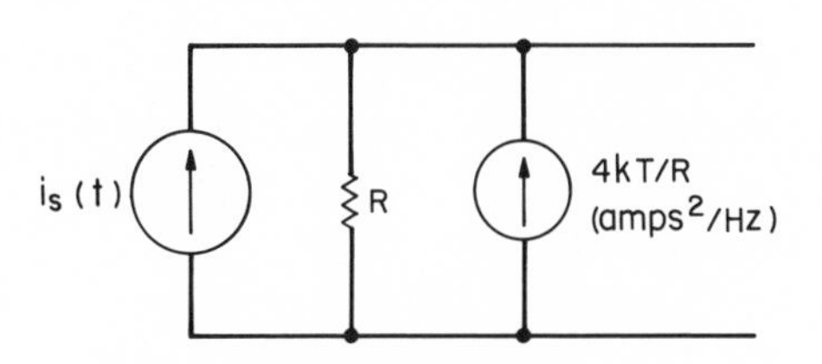

Figure 4-16 Resistive Signal Source.

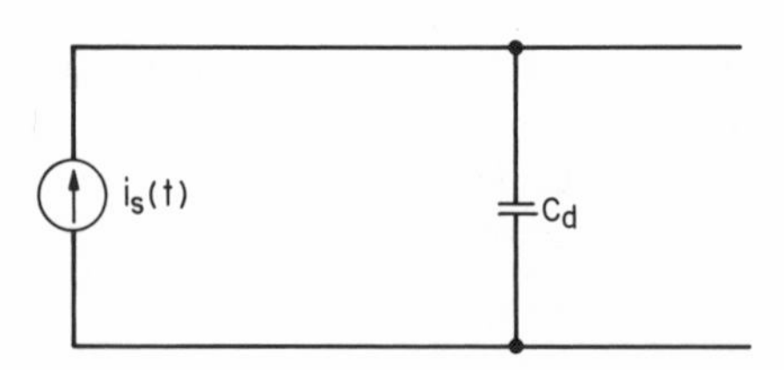

Figure 4-17 Capacitive Signal Source.

A photodiode is often referred to as a capacitive source. Other examples of capacitive sources are vidicon tubes and biological signal sources. When we design a preamplifier to work with a resistive source such as the one shown in Figure 4-16 we characterize the noise added by the preamplifier electronics by a measure called the noise figure. The noise figure is the ratio of total noise at the preamplifier output to the portion of the output noise due to the intrinsic noise of the signal source itself. If the preamplifier adds a noise which is small compared to the intrinsic noise of the signal source, then the noise figure is close to unity. If the noise figure is close to unity, there is little benefit in trying to further reduce the preamplifier noise contribution.

When we have a capacitive source as shown in Figure 4-17 there is no meaningful intrinsic noise against which to compare the preamplifier noise. The noise figure in this case is always very large, since the preamplifier noise dominates the total. For a capacitive source, a more meaningful measure of preamplifier noise can be obtained by asking whether or not individual hole–electron pairs generated in the detector produce an observable signal at the preamplifier output. If the individual hole–electron pair responses are observable, than the receiver performance will be limited by fundamental quantum effects.

We can define a parameter, Z, as follows:

$$Z = \frac{\text{output rms noise}}{\text{output response to an electron–hole pair}} = \sigma_o / \nu_e \tag{4-10}$$

where σ_o is the rms noise at the receiver output in the band of frequencies occupied by the modulation of the input signal, and where ν_e is the peak response produced at the receiver output by an individual hole–electron pair generated in the detector.

If Z is a number less than unity, then individual pair responses are observable. If Z is larger than unity, then only the cumulative output due to many generated pairs is observable. Consider the following example.

Figure 4-18 shows a simple optical receiver where a photodetector is coupled to a conventional amplifier modeled as shown. In this example we assume that the conventional amplifier has a physical 50-Ω resistor at its input, and that the resistor produces Johnson (thermal) noise of spectral density $4kT/R$ (A^2/Hz). The receiver has a bandlimiting filter of bandwidth B. Presumably B is the bandwidth required to accommodate the modulation of the incoming light signal. To calculate Z we must look at the output of the receiver. We must calculate the rms noise due to the input resistance of the amplifier; and we must calculate the response at the same point produced by a pair generated in the detector. Using standard noise theory we obtain the variance of the noise at the receiver output as follows:

$$\sigma_o^2 = 4kT50A^2B \tag{4-11}$$

where A is the amplifier transconductance (G_m) × the filter gain and $kT = 4 \times 10^{-21}$ J (at room temperature).

We can derive the amplitude of the output pulse produced by a pair generated in the detector as follows. The generated pair produces a displacement current whose area is e, the charge of an electron. This current flows through the 50-Ω resistor (we neglect capacitance in this example) to produce an input voltage waveform whose area is $50 \times e$ (V-s). This results in an output waveform whose area is $50 \times e \times A$ (V-s). The bandlimiting filter has bandwidth B. Thus the pulse produced at the filter output has a duration of roughly $1/B$. Thus the peak of the response produced at the filter output is roughly $50 \times e \times A \times B$ (V).

We then obtain the following expression for Z:

$$Z = \frac{(4kT50A^2B)^{1/2}}{50eAB} = \left[\frac{4kT}{50e^2B}\right]^{1/2} = 10^8\left[\frac{1}{B}\right]^{1/2} \tag{4-12}$$

We see that Z decreases as the bandwidth B increases. This is because the peak response grows linearly with B, while the rms noise grows only as

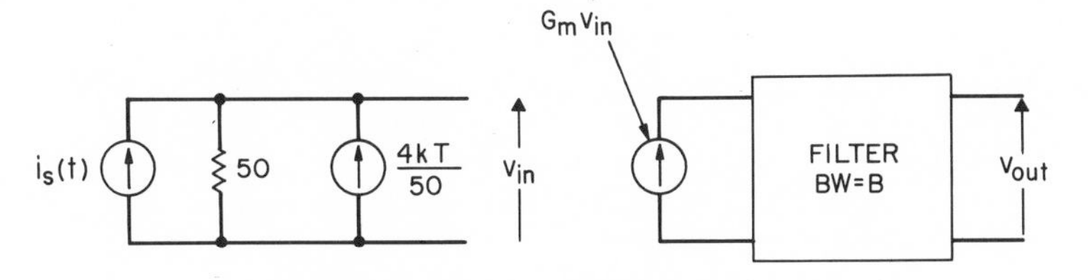

Figure 4-18 50-Ω Amplifier Example.

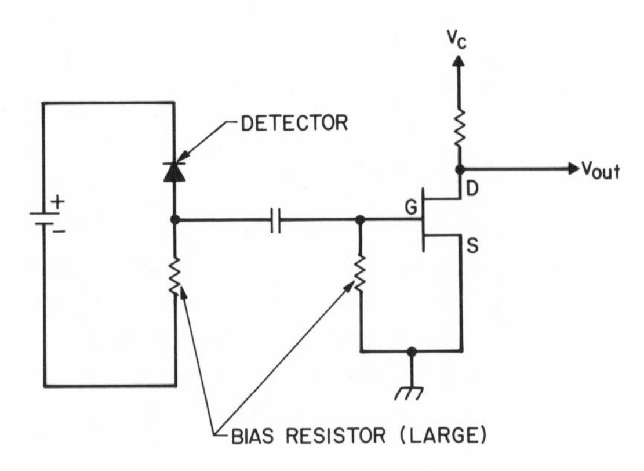

Figure 4-19 FET Preamplifier Example.

the square root of B. However, at very high bandwidths other noise sources which we have neglected (due to the preamplifier active components) become important and cause Z to increase with bandwidth. As an example, if the bandwidth B is 10^8 Hz, then Z is approximately 10,000. This implies that 10,000 pairs must be generated in the detector to produce a cumulative response which is just equal to the noise at the preamplifier output.

With careful electrical and physical design (see also Section 7.3, Figure 7-20) we can implement preamplifiers with values of Z of between several hundred and several thousand, depending upon the bandwidth and other tradeoffs. Figure 4-19 shows a preamplifier based on a field-effect transistor. Figure 4-20 shows an equivalent circuit for this simple preamplifier. The input of the field effect transistor (FET) is a capacitance. The transistor produces an output current proportional to the input voltage, with proportionality constant G_m. There is a noise source in series with the input voltage produced across the parallel detector and FET capacitances, associated with the thermal noise of the source-drain channel. This noise source has single-sided spectral density $2.8kT/G_m$ (V^2/Hz).

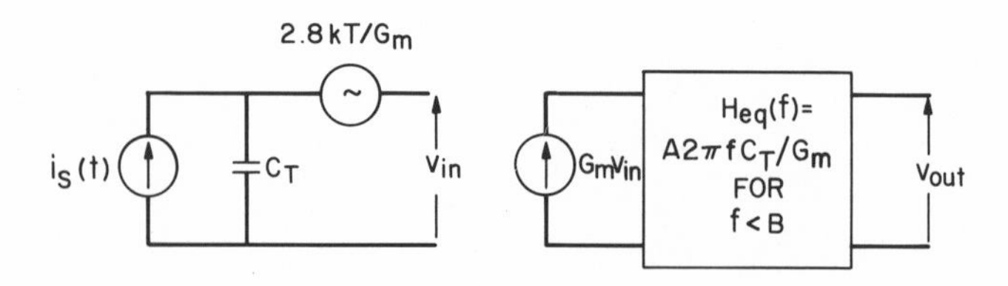

Figure 4-20 FET Preamplifier Equivalent Circuit.

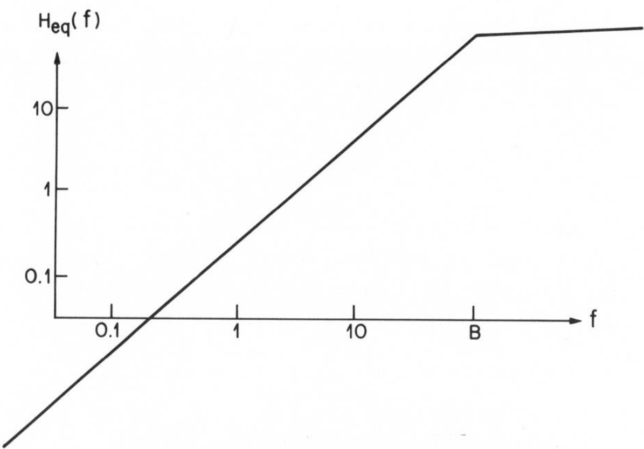

Figure 4-21 Equalizer Frequency Response.

The parallel capacitance at the input of the FET causes the signal current produced in the detector to be integrated. This is compensated by a differentiating network at the output of the FET as shown in Figures 4-21 and 4-22, resulting in a net frequency response which is flat out to some desired bandwidth B. Note that the preamplifier input noise source is not integrated at the input. Thus it produces a noise at the receiver output which does not have a flat spectrum. We can next calculate Z for a typical FET preamplifier using the following parameter values. We shall assume that the detector capacitance is 0.5 pf and that the preamplifier input capacitance plus any stray capacitance at the input totals 0.5 pf. Thus the net input capacitance, C_T, totals 1.0 pf. We shall assume that the FET is a GaAs device with a transconductance G_m of 10 millisiemens (mS). The rms noise at the receiver output can be obtained from standard

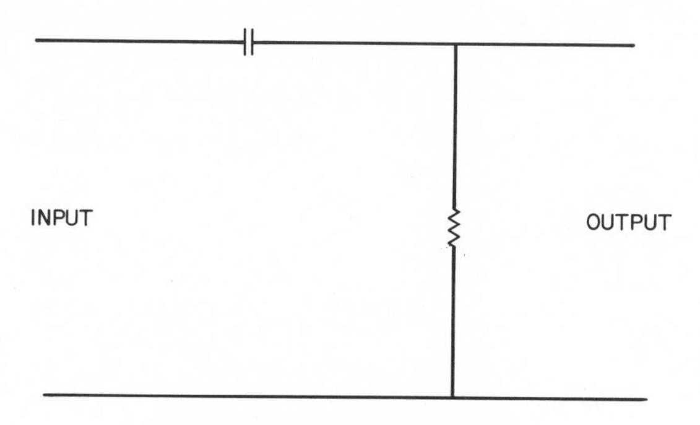

Figure 4-22 Equalizer Schematic.

noise theory by integrating the frequency dependent noise produced by the input noise source over the band of frequencies B. We obtain the following result:

$$\sigma_o^2 = \frac{2.8kT}{3G_m} (2\pi C_T)^2 A^2 B^3 \tag{4-13}$$

The peak of the response produced at the receiver output by a pair generated in the detector can be calculated (as we did in the 50-Ω preamplifier case above) to be

$$v_e = eAB \tag{4-14}$$

Thus we obtain the following expression for Z:

$$Z = \frac{\sigma_o}{v_e} = \left[\frac{2.8kTB}{3G_m e^2} (2\pi C_T)^2 \right]^{1/2} = 2.4 \times 10^{-2} B^{1/2} \tag{4-15}$$

Note that Z increases as B increases due to the input noise source of the transistor (as foreshadowed above). Note also that Z is proportional to $C_T/G_m^{1/2}$. Thus to minimize Z we wish to minimize the total input capacitance while at the same time using a device with a large value of G_m. If we set B=10^8 Hz and if we use the parameter values given above then we obtain a value of Z equal to 240.

We can also make low noise preamplifiers for capacitive sources using bipolar transistors. One can show that with bipolar transistor preamplifiers the value of Z for a given selection of transistors is independent of B over a wide range of frequencies, provided the preamplifier is optimally biased for the value of B selected. A typical value of Z for a

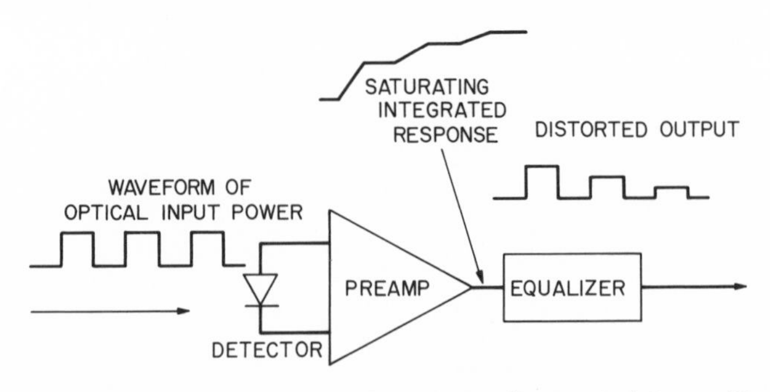

Figure 4-23 Nonlinear Effects in High Impedance (Integrating) Preamplifiers.

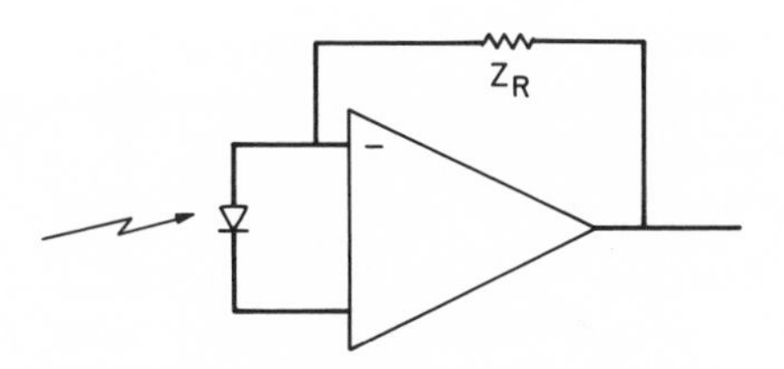

Figure 4-24 Transimpedance Preamplifier.

bipolar transistor preamplifier in the range of bandwidths from 100 MHz to 2 GHz is around 1000, provided appropriate transistors are used.

Preamplifiers which are designed to minimize Z often have very limited dynamic range. If we reexamine the FET preamplifier described above we observe that the compensation network at the output of the preamplifier can only remove the integration caused at the preamplifier input if the preamplifier operates linearly. Low frequencies in the input light signal modulation result in low frequency currents in the detector response which can be integrated by the total capacitance at the preamplifier input to produce large voltages. Thus if the optical signal is above its minimal detectable level, and if the modulation contains low frequency components, the preamplifier may overload (become nonlinear). Figure 4-23 illustrates how a sequence of pulses at the input to the

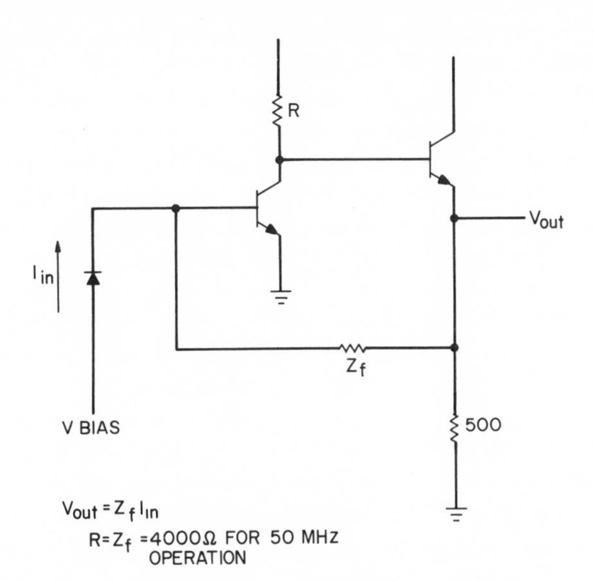

Figure 4-25 Example of a Transimpedance Preamplifier.

preamplifier might be distorted by this process. To increase the dynamic range of the preamplifier, several aproaches have been used. Figure 4-24 shows a transimpedance preamplifier design, where the preamplifier is used as an operational amplifier. In order for the transimpedance design to work properly, the gain around the feedback path (loop) must be large compared to unity. Since the feedback resistance and the amplifier input impedance form a voltage divider, the allowable value of the feedback resistance is limited. The feedback resistance adds noise at the amplifier input which typically dominates the noises from the active components (transistors) in the preamplifier. Thus a transimpedance preamplifier typically has more noise than an integrating design made from the same components. A useful rule of thumb is that Z is 2–3 times larger for the transimpedance design (vs the integrating design). Figure 4-25 shows a transimpedance preamplifier using bipolar transistors and having a bandwidth of about 100 MHz. The Z value for this preamplifier is about 3000.

Other approaches to improving the dynamic range of a preamplifier (beyond the use of a transimpedance design) include the use of nonlinear elements at the input of the preamplifier (to prevent overload of the preamplifier itself) and the use of an adjustable transimpedance (to reduce the transimpedance value at high signal levels).

4.2.2 APD Control Circuitry

The average multiplication (gain) of an avalanche photodiode is dependent upon the applied reverse bias and the temperature of the device. Figure 4-4 shows curves of responsivity vs voltage and temperature for a typical APD. In a practical system application it is usually necessary to provide some mechanism for assuring that the APD gain is correctly set. When relatively low values of gain are used (i.e., where the temperature sensitivity is modest) open loop compensation of the applied reverse bias can be sufficient. Thus for those situations one might implement a controller for the high voltage APD supply whose temperature characteristic approximately tracks the requirements for maintaining nearly constant APD gain. When higher values of APD gain are used, closed-loop control mechanisms may be necessary (unless the APD is maintained in a controlled ambient). Figure 4-26 shows a receiver design often implemented in digital systems where the APD is included in the automatic gain control loop of the receiver.

The output level of the receiver can be adjusted via either the gain of the APD or the gain of the variable gain amplifier (VGA). The output peak-to-peak level is measured by a peak-to-peak detector (a nontrivial circuit at high bandwidths), averaged, and compared to a reference. The low frequency control circuitry of the AGC system produces an error voltage in

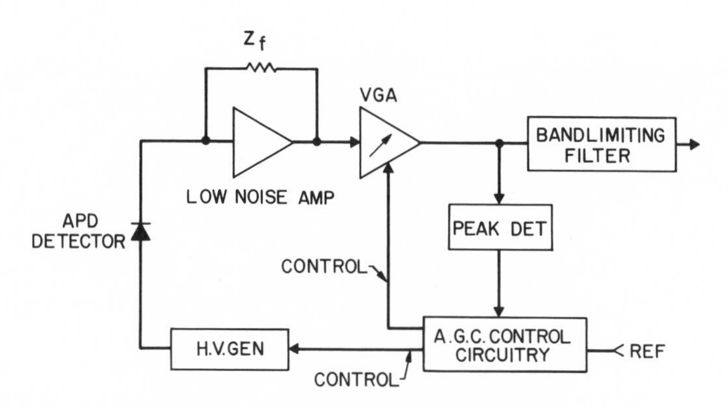

Figure 4-26 Feedback Control of APD Gain.

response to the difference of the reference and the averaged peak-to-peak measurement. This error voltage is used to adjust either of the available gain mechanisms (or both). One possible strategy is to have the APD high voltage supply controller respond over a range of AGC error voltages which is separate from the range of error voltages which control the VGA. For example, the VGA could be adjusted to be at full gain for error voltages between 0 and 6 V, and to be at reduced gains for error voltages between 6 and 12 V. Meanwhile the high voltage controller could be adjusted to produce some minimum level of output voltage (required for proper APD operation) for error voltages above 5 V, but to produce higher voltages for error voltages between 0 and 5 V. In this scheme, if the output of the receiver is too high, the AGC control loop will first reduce the APD gain down to some minimum permissible value (set by the requirement that the *i*-region in the device be swept out), and will then reduce the gain of the VGA. This scheme will automatically adjust the gain of the APD to compensate for temperature changes when the APD is operating in the high range of gains (where the AGC is controlling its voltage). When the APD voltage is fixed at its lower limit, the APD will be operating at low gains where the precise gain value is not critical and where the APD gain is less sensitive to temperature changes. Figure 4-27 shows an example of a high voltage controller circuit which uses a shunt current path to drop the voltage applied to the APD.

The Zener diodes limit the maximum and minimum APD voltages. There are some cautions which must be observed in using this design. The APD will not operate as a high speed device unless it is biased with sufficient voltage to deplete its *i*-region. The required voltage may vary amongst APDs which are nominally the same. Thus some specification needs to be made regarding the voltage required for *i*-region depletion, and

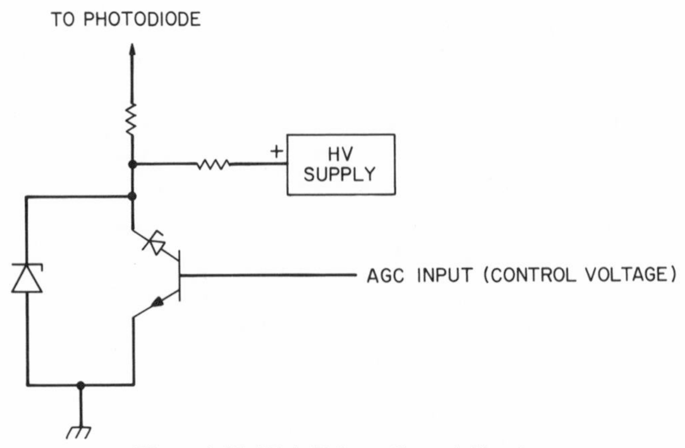

Figure 4-27 High Voltage Control Circuit.

the lower limit of the bias voltage must be set accordingly. The voltage required to produce the desired gain in one APD may produce breakdown in another nominally similar device. In some applications caution must be exercised to limit the reverse bias to a value below breakdown (although this is not always a problem). In designing the feedback loop one must be careful that the differing time constants of the multiple gain control mechanisms do not result in oscillatory behavior. It can also be shown that if the control loop is set up initially to adjust the APD gain to the optimal value at some minimum anticipated optical signal level, then the gain will not be optimal (too low) at higher signal levels. (The loop is keeping the product of received power level and avalanche gain constant.) It can be argued that when the optical signal level increases, the receiver performance increases, even with the suboptimal APD gain setting. However, in some applications this may not be sufficient.

In analog systems, the required avalanche gain is usually low enough (as we shall see in Section 4.2.3 below) that the temperature sensitivity of the APD gain can be ignored or tracked by open loop methods. However, this would not be the case if the absolute end-to-end gain of the link is specified to tight tolerances. In such a situation some sort of pilot tone or other measure of APD gain must be available at the receiver output, in order to set the gain by closed loop control.

4.2.3 Receiver Performance [8, 9, 14, 16, 17]

The performance of a receiver is usually measured by the minimum required optical power level to obtain a given fidelity of extraction of the information contained in the modulation of the optical signal. Other measures of performance include dynamic range, sensitivity to power supply voltage deviations from nominal and power supply noise, sensitivity to electromagnetic interference, linearity, gain stability, transient and overvoltage protection, etc. Many of these concerns involve classical circuit design considerations and are beyond the scope of this book. We shall focus on the receiver sensitivity.

In a digital system, receiver performance is measured in terms of bit errors caused by random noise. (We exclude in this discussion bit errors caused by power supply transients and interference from other circuitry.) In classical communication and sensing systems the noise which accompanies the signal is often (but not always) additive and Gaussian in statistics. As a result, the bit-error rate can be predicted from the ratio of peak signal-to-rms noise at the input to the digital comparator (decision circuit which compares its input to a threshold, once per bit interval). In optical systems using *p-i-n* detectors, the additive Gaussian noise model is very accurate, and classical projections of the bit error rate from the signal-to-noise ratio are valid. In optical systems using APD detectors the signal-dependent quantum noise, enhanced by the random gain, often dominates. This noise is neither simply additive nor Gaussian in statistics. Approximations for the bit error rate can be made my calculating signal-to-noise ratios and ignoring the non-Gaussian nature (using the formulas based on additive Gaussian noise) of the noise. Alternatively complex analyses made on more exact models of the noise can be performed. Furthermore, more exact analyses can be used to calibrate approximate results based on the Gaussian model. One typically finds that predictions of the required power at the input of the optical receiver to produce a desired bit error rate (the receiver sensitivity) based on the Gaussian approximation are accurate to within about 1 or 2 dB of the values predicted by more exact analyses and of the values measured in experiments. When a *p-i-n* detector is used, the required received energy per optical pulse to achieve a bit error rate of 10^{-9} is $12 \times Z \times hf$, where hf is the energy in a photon and Z is the preamplifier noise parameter. For $Z = 1000$ we require 12000 photons per pulse, compared to the quantum limit of 21 photons per pulse. Thus a practical receiver with a *p-i-n* detector is 27 dB less sensitive (typically) than an ideal receiver. If we use avalanche gain, we obtain an improvement which depends on the statistics of the gain mechanism. This, in turn, depends upon the material from which the APD is fabricated. More specifically, the obtainable improvement depends upon the ionization ratio (k) for holes and electrons. For good silicon detectors (k less than 0.05) receiver sensitivity improvements of 10–15 dB are typically obtainable (relative to a *p-i-n* detector) at optimal avalanche gains of between 50 and 150. (Note: if the detector had deterministic gain, then a gain of 100 would result in a 20-dB improvement in receiver sensitivity.) Thus with silicon detectors one can obtain performance within about 12 dB of the quantum limit. (Even better results can be obtained with heroic efforts.) The statistics of germanium and InGaAsP detectors are not as good as those of silicon (although efforts continue to obtain improved devices). With these long wavelength detectors performance improvements of 5–7 dB are typical relative to the *p-i-n* device and are obtained at gains of about 10–20. Figure 4-28 gives guidelines for the performance that can be expected from typical receivers using on–off modulation, and over a range of data rates. These curves assume that the receiver is adjusted properly to the incoming optical power level and that the decision threshold and sampling times are correct. In some digital communication applications, such as data buses, short bursts of digital symbols (packets) are used, which do not provide long periods for the receiver to adjust its level control and timing recovery circuitry. In those applications the receiver performance may be somewhat poorer than what is shown in the curves of Figure 4-28.

In practical systems, the bit error rate can also be influenced (or limited) by interference between pulses (caused by inadequate bandwidth),

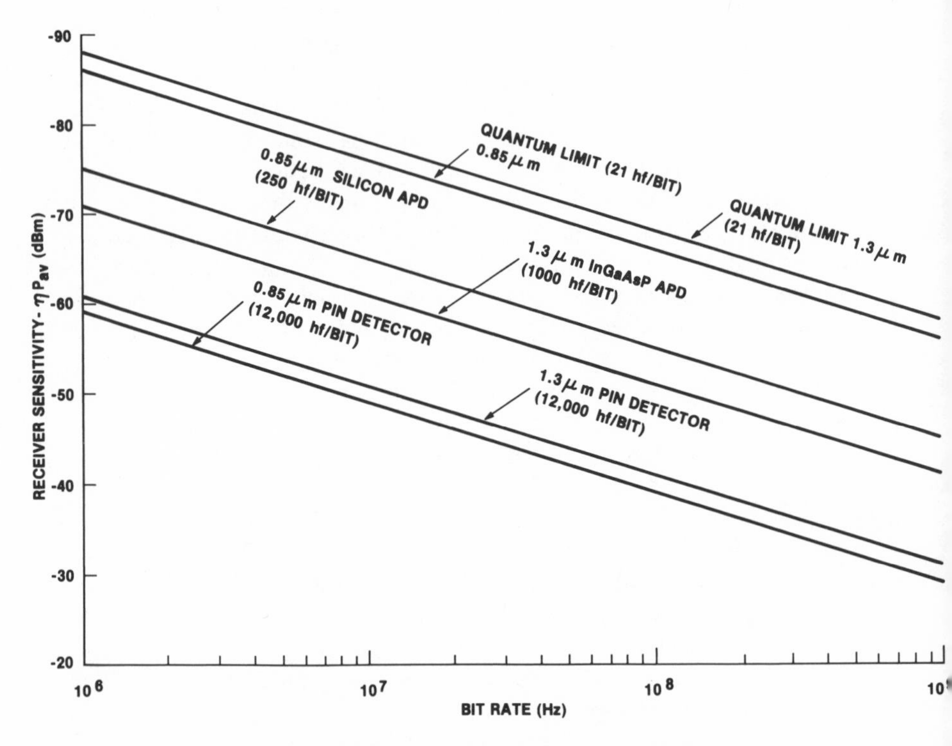

Figure 4-28 Typical Receiver Performance vs Bit Rate.

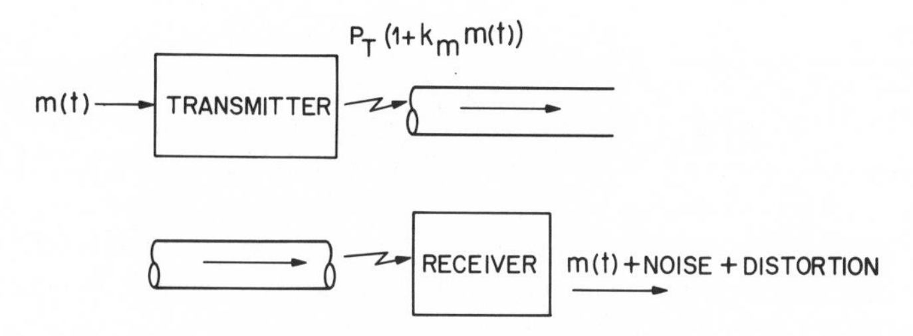

Figure 4-29 Simple Analog System.

nonlinear effects in the transmitter, transmitter fluctuations, and modal noise and mode partition noise phenomena (discussed in Chapter 6 below), amongst other things. We shall discuss these practical problems further in Chapter 7.

The performance of an analog communication system is typically measured by the ratio of desired signal (peak or rms) to noise at the output of the system. If the system uses a modulation technique such as fm, then the signal-to-noise ratio is measured at the output of the fm demodulator, or inferred from the signal-to-noise ratio at the input to the demodulator (using theoretical relationships between the two). In analog systems, particularly with multiplexed signals, linearity can also be critical. Linearity and noise trade off against each other in most situations.

Consider the simple analog system shown in Figure 4-29. An optical source is biased to some average level of output P_o and modulated above and below this level by a message $m(t)$ having zero average value and unity peak value. The modulated waveform produced by the optical transmitter is

$$p(t) = P_o\,[1 + k_m\, m(t)] \tag{4-16}$$

In equation (4-16) $m(t)$ is assumed to be limited to a peak value of unity. The parameter k_m is called the modulation index. Since the optical power must be a positive quantity, we require k_m to be less than 1.0. An attenuated version of the transmitted signal arrives at the receiver. The output of the receiver consists of a scaled version of $m(t)$ plus noise. The noise arises from the quantum noise produced in the detection process (enhanced by avalanche gain, if used), thermal noise from the preamplifier, and miscellaneous (possibly dominating) noise resulting from intrinsic transmitter fluctuations, and modal noise effects (to be discussed in Section 6.1). We shall ignore these miscellaneous noises temporarily. The general form of the peak signal-to-rms noise ratio at the output of the receiver is given by

$$\text{SNR} = \frac{(\eta\, P_r\, k_m/hf)^2}{[2\eta\, P_r\, F(G)B/hf] + (Z^2B^2/G^2)} \tag{4-17}$$

where hf is the energy in a photon at the wavelength of operation, P_r is the average received power level, η is the detector quantum efficiency, G is the avalanche gain and $F(G)$ is the avalanche gain excess noise factor (if avalanche gain is used), k_m is the modulation index, Z is the receiver noise parameter, and B is the bandwidth being used [required to accommodate $m(t)$]. We can make a few observations. For a given average optical power level, there is an optimal value of avalanche gain, which maximizes the signal-to-noise ratio. As the desired SNR increases, the required value of average optical power increases. Furthermore, as the desired SNR increases, the value of optimal gain decreases. The SNR is proportional to the square of the modulation index. (Use of a higher modulation index will reduce the linearity, in general, as will be discussed further in Chapter 10.) Figure 4-30 shows curves of the required optical power to achieve the signal-to-noise ratios: 1, 100, 10^4, and 10^6; vs the avalanche gain, for the following sample parameter values: quantum efficiency equal to 0.7, $hf = 2 \times 10^{-19}$(J), $F(G) = 2 + 0.04G$, $k_m = 0.5$, $B = 20$ MHz, $Z = 1000$.

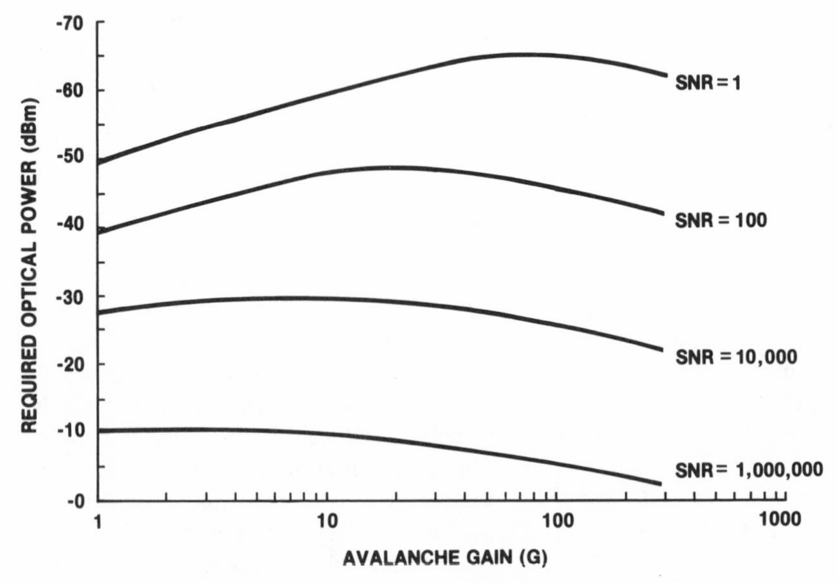

Figure 4-30 Required Power vs SNR and Avalanche Gain.

We see from these curves that the optimal avalanche gain is near unity for signal to noise ratios above 10^6. If in equation (4-17) we assume that Z is fixed, independent of B, and if we define a parameter $Y = P_r / B$, then we obtain a "universal" set of curves which gives the required value of Y to achieve a desired SNR, independent of B.

We can obtain higher signal-to-noise ratios at the link output, for a given received optical power level, by using bandwidth expansion techniques, such as fm. This follows from classical communication theory. Techniques such as fm have the additional advantage of being less susceptible (or immune) to nonlinearities in the transmitter and the receiver. Furthermore, intrinsic transmitter noise and other effects can limit the attainable signal-to-noise ratio to values below those predicted in Figure 4-30. Modulation techniques like fm can produce a signal-to-noise ratio at the output of the demodulator which is substantially higher than the SNR at the input to the demodulator (beyond the improvement attributable to the smaller bandwidth at the demodulator output). We will discuss analog systems further in Chapter 10 below.

Problems

1. A nonmultiplying (p-i-n) detector produces a 5 μA current in response to a 6 μW optical illumination. The optical wavelength is 1.3 μm. What is the quantum efficiency of the detector? What is the responsivity of the detector?

2. An ideal nonmultiplying p-i-n detector produces a Poisson-distributed number of electron–hole pairs in response to illumination by an optical pulse of energy E_R (joules). The number of pairs produced on the average (mean of the Poisson distribution) is equal to the number of photons corresponding to E_R (joules).

   Assume that the light wavelength is 1.3 μm. Assume that the detector absorbs all of the incident light (100% quantum efficiency). Assume that the amplifier attached to the detector is so ideal, that you can count the number of hole–electron pairs produced by the detector (exactly). Assume that there is no mechanism to produce hole–electron pairs in the detector other than the received pulse of light energy (no dark current).

   We wish to communicate one bit of information to the receiver by sending a light pulse of energy given by 0.1 E_R (off) or 1.0 E_R (on). That is when we send a "zero," the optical transmitter is not perfectly off, thus the received energy is 10% of what is received when we send a "one." We decide whether a zero or a one has been sent by comparing the number of hole–electron pair produced to a fixed threshold value. That is, if we receive more than a certain number of pairs, we decide that at "one" was sent.

If we want the probability of error to be 10^{-9} for mistakenly deciding a one was transmitted when a zero was actually transmitted and vice versa, what is the required threshold (in detected pairs) level, and how big must E_R be (in photons). (You can solve this problem using a small computer or you can approximate the Poisson distributions by Gaussian distributions. That is, the probability that a realization of a Poisson random variable will *deviate from* its mean by six times its standard deviation is approximately 10^{-9}. The standard deviation of a Poisson random variable is the square root of its mean.)

3. An optical receiver consists of a detector and a preamplifier to amplify the current generated by the detector in response to incoming light. A particular receiver has a preamplifier with a bandwidth of $0–5 \times 10^7$ Hz. The preamplifier adds noise which is equivalent to the Johnson noise of a 4000-Ω resistor acting as a current noise generator in parallel with the detector equivalent signal current generator (the noise of a 4000-Ω resistor has spectral density $4kT/4000\ A^2/\text{Hz}$, where $4kT = 1.6 \times 10^{-20}$ J at room temperature.)

   Using the same approximations as in the text, calculate the value of Z (output noise standard deviation/response at the output produced by a single hole–electron pair emitted by the detector) for this preamplifier. (That is, approximate the response at the output produced by an emitted electron–hole pair as rectangular in shape with duration 1/(50 MHz) and area $e \times$ the receiver gain in volts/ampere.)

4. In order to achieve a bit error rate of 10^{-9}, a particular optical receiver requires a minimum input level of 3000 photons per "on" pulse. Assume that digital ones (on pulses) therefore, correspond to 3000 photons of received energy. Consistent with this, the receiver accommodates a 10% extinction ratio. Therefore, assume that in the "off" condition (digital zero sent) the received energy is 300 photons. If the wavelength is 1300 nm and the data rate being communicated is 100 Mb/s (half ones and half zeros on the average), what is that average power required at the receiver in watts?

5

Optical Components

In the previous sections we have described optical sources, detectors, fibers, and connection components. These are required building blocks for any fiber optic system. In some systems additional optical components such as access couplers, power dividers, wavelength multiplexers, external modulators, and optical switches are also needed.

5.1 Passive Components

Passive components used in optical systems (other than splices and demountable connectors) include access couplers (taps), power dividers and star couplers, wavelength multiplexers, and mechanical optical switches.

5.1.1 Access Couplers

The purpose of an access coupler is to allow a fraction of the light propagating through an optical fiber to be removed (tapped out) or to allow optical power to be added at an intermediate point along a fiber link. Figure 5-1 shows a "brute force" method of implementing an access coupler. Light propagating in a fiber is collimated (magnified into a larger diameter roughly parallel beam) by a lens and directed toward a partially reflecting mirror. Part of the light passes through the mirror and is focused by a second lens into an outgoing fiber. Part of the light is reflected toward a third lens which focuses that light into another outgoing fiber (or on to a local detector). A fourth lens allows light from a second incoming fiber to be added to the first outgoing fiber. Thus we can tap out some light and add in other light. Although such a coupler is not generally very practical (it is difficult to align and mechanically bulky and fragile) it illustrates some important points common to all passive access couplers. For each electromagnetic mode (field pattern) incident upon the coupler, the coupler obeys the principle of reciprocity. That is, if the coupler

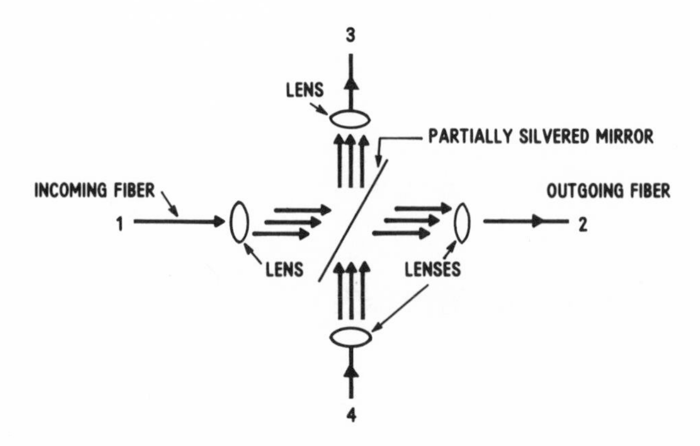

Figure 5-1 Access Coupler Implemented with Bulk Optics Components.

removes 10% of the light power passing left to right (directs 10% of the power from port 1 to port 3), then it will only couple 10% of the power incident from the local source (port 4) into the outgoing fiber (port 2). Reciprocity need only apply on a mode-by-mode basis. For example the coupler could remove 10% of the power in one mode (direct it from port 1 to port 3) and couple in 90% of the power in a different spatial mode (direct it from port 4 to port 2). The fact that a coupler could have a mode-dependent coupling ratio can be advantageous or detrimental, depending upon the application. Note also that for single mode fibers there is only one spatial mode to work with (on the incoming and outgoing fibers at ports 1 and 2) and therefore reciprocity must hold (unless the coupler is polarization dependent). Figure 5-2 illustrates a design for an access coupler where a partially reflecting mirror is formed on the polished ends

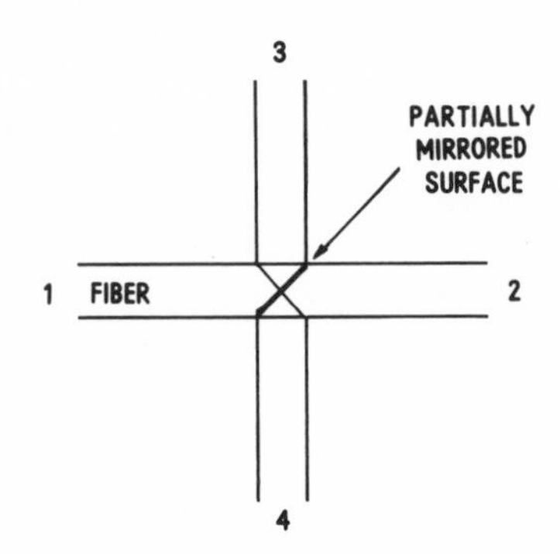

Figure 5-2 Access Coupler Implemented with Silvered Fiber Ends.

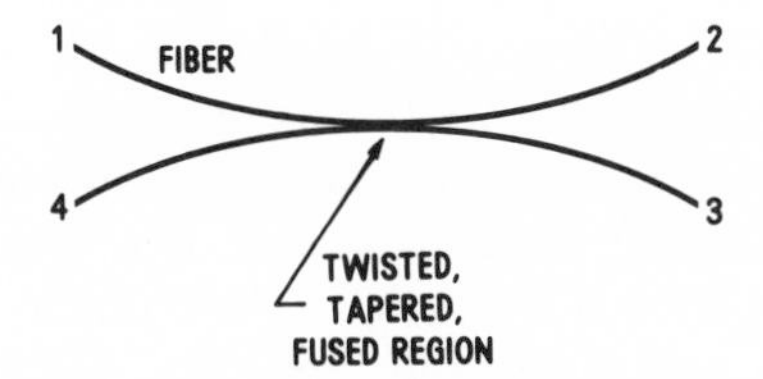

Figure 5-3 Fused Tapered Access Coupler.

of four fibers joined in the shape of a cross. This design could be manufactured using automated polishing and assembly techniques, and is more rugged than the design shown in Figure 5-1. However, in this design, the the light incident upon the mirror has a relatively large spread in angles (within the critical angle of the fiber), resulting an a reflectivity which is somewhat mode sensitive. (The reflectivity of a partially silvered mirror depends upon the incident angle and the polarization.) Figure 5-3 shows a popular access coupler used for multimode fibers. Two fibers are twisted and the four ends are placed under tension. The twisted region is then heated so that the fibers taper and fuse together. Light entering port 1 (for example) leaks into the cladding when it passes into the tapered region. (A smaller diameter fiber supports fewer modes.) This light then mixes with the cladding modes of the other fiber. Remarkably, as the light leaves the fused tapered region it is recaptured by the cores of the two outgoing fibers (2 and 3 in this case) with very little light scattered out of the coupler (lost). How much light couples between fiber 1-2 and fiber 4-3 depends upon the details of the fused region and upon the distribution of the power amongst the modes of the incoming fiber. With careful control of

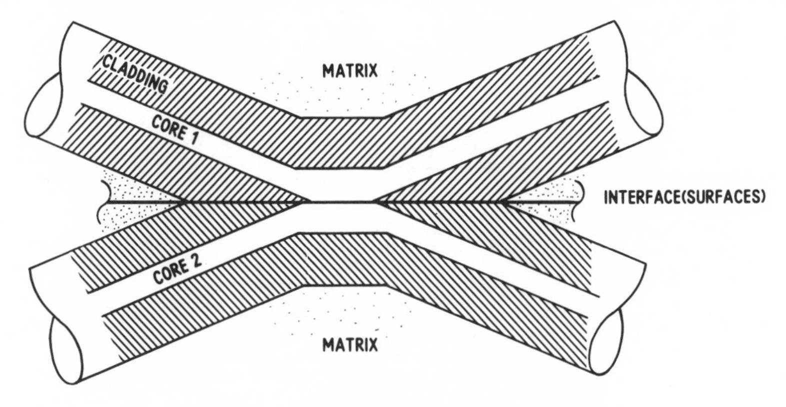

Figure 5-4 Single Mode Fiber Access Coupler.

the fabrication process couplers with coupling ratios between 1%–99% and 50%–50% can be produced (where the coupling ratio is defined based upon some assumed distribution of power amongst the modes of the incoming fiber). Figure 5-4 shows a coupler which has been reported for single mode fibers. The fibers are set in blocks which are carefully polished to remove the cladding and expose the cores. The blocks are then placed in contact to allow the cores to interact.

5.1.2 Wavelength Multiplexers

In some system applications it is desirable to have two or more wavelengths propagating within the same fiber. Wavelength selective devices are needed to allow these optical signals to be used independently. Figure 5-5 shows an example of a wavelength multiplexer for two wavelengths fabricated from miniature lenses and a wavelength selective (dichroic) mirror. At port 1 two wavelengths arrive on the incoming fiber. A lens collimates the incoming beam and directs it toward the mirror. Wavelength 1 is mostly transmitted, while wavelength 2 is mostly reflected. The angled mirror causes wavelength 2 to be focused to a spot and captured by the fiber at port 3. Light at wavelength 1 is focused by a second lens to a spot captured by the fiber at port 2. This device also works in reverse. Light at wavelength 1 coupled into port 2 will come out at port 1. Light at wavelength 2 coupled in at port 3 will also come out at port 1. The device can be implemented with both single mode and multimode fibers. Excess losses of less than a 3 dB have been reported for back-to-back multiplexer pairs. The selectivity of the mirror in the transmitting direction (rejection of the unwanted signal) can be very high (30 dB or more). However, the selectivity of the mirror in the reflecting direction is relatively low because some unwanted light will always be reflected. Increased selectivity in this direction can be obtained by using a supplementary optical filter which transmits the desired wavelength.

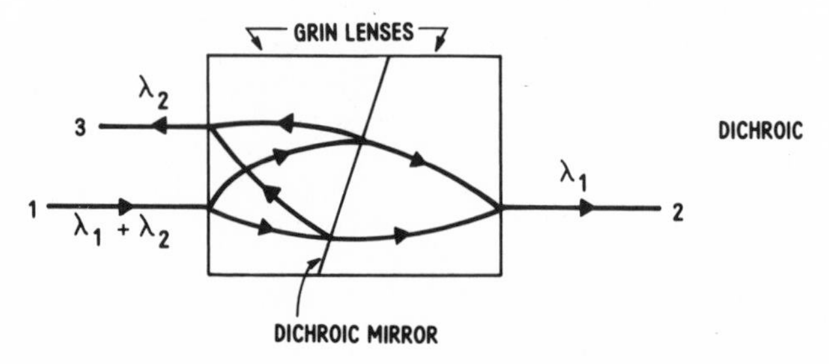

Figure 5-5 Miniature Bulk Optics Wavelength Multiplexer with a Dichroic Filter.

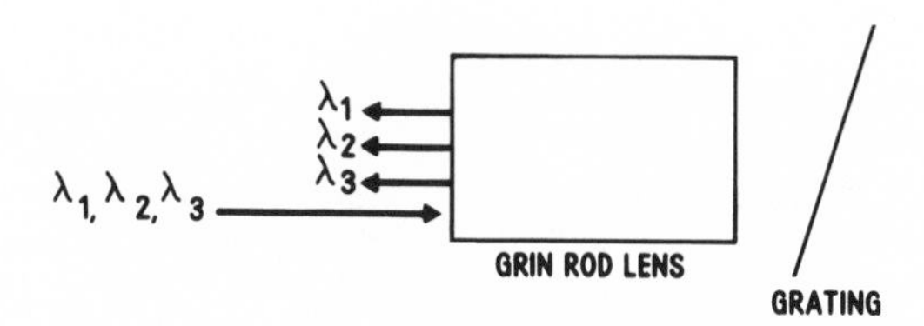

Figure 5-6 Miniature Bulk Optics Wavelength Multiplexer with a Grating.

Figure 5-6 shows a three-wavelength coupler which uses an optical grating (closely spaced lines ruled on a substrate) which acts as a wavelength selective mirror. Different wavelengths are reflected at different angles. Incident light is collimated by the lens and directed to the mirror. The three wavelengths are reflected at different angles and focused by the lens back into separate spots for coupling into the three outgoing fibers. This device also works in reverse, coupling three wavelengths into a single fiber. Although wavelength multiplexers using dichroic mirrors and gratings have been demonstrated in the lab (and used in some experimental fielded systems) one should keep in mind that the alignment tolerances and needed parameter controls (on the mirrors) are very tight. These problems are enhanced when closely spaced wavelengths must be multiplexed. Degradations due to aging and temperature effects can be serious. Therefore, although wavelength multiplexing is an emerging capability which will be used in practical systems, each application must be carefully analyzed to match the available technology with the requirements.

One can also fabricate wavelength selective couplers with integrated optics technology as shown in Figure 5-7. Here waveguides are formed in the surface of a glass substrate (or other transparent material) using diffusion through masks formed by photolithographic techniques. The waveguides are regions of higher index of refraction than the surrounding substrate. If the waveguides run parallel and are closely spaced (within a few micrometers) they can couple energy. Furthermore, the coupling can be made to be wavelength sensitive by careful control of the waveguide geometry. Thus we can arrange to have a monolithic device which accepts

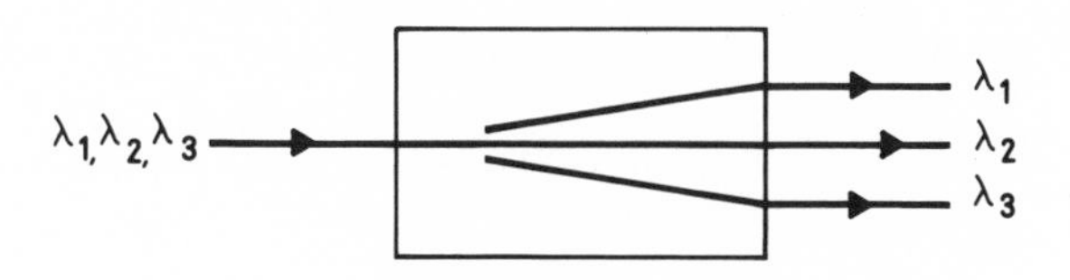

Figure 5-7 Integrated Optics Wavelength Multiplexer.

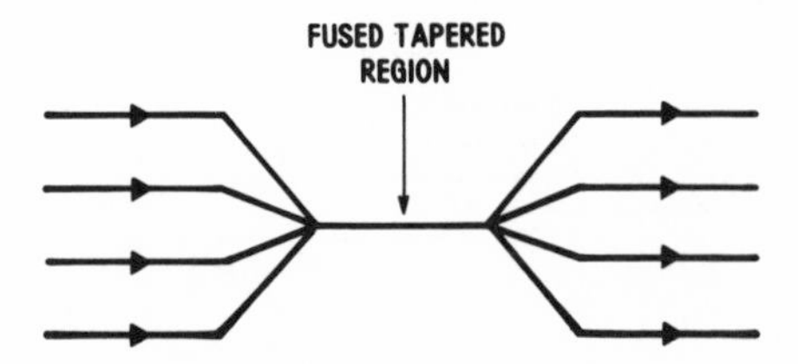

Figure 5-8 $N \times N$ Star Coupler.

three wavelengths on the incoming port and produces separated wavelengths at the outgoing ports. Integrated optical devices are still in a research phase of emergence (for the most part). Problems currently being addressed include tight control of the tolerances of the fabricated waveguides, efficient coupling from fibers to integrated optics surface waveguides and from the surface waveguides to the fibers, and polarization sensitivity of integrated optics devices. Integrated optics devices generally work only with single mode light signals.

5.1.3 Power Dividers and Star Couplers

A power divider splits an incoming optical signal into several outgoing signals. A star coupler is a power divider which has several input ports. These devices are extensions of the concept of an access coupler discussed in Section 5.1.1 above. (An access coupler is a two-input/two-output star coupler and can work as a power divider if only one input port is used.) Figure 5-8 shows an $n \times n$ (n-input/n-output) star coupler which can also be used as a $1 \times n$ power divider. It is fabricated by the same techniques as the tapered fused access coupler described in Section 5.1.1 above. Star couplers with up to 64 inputs and 64 outputs fabricated using the fused-tapered approach are commercially available. When the number of inputs and outputs is large, they are fabricated in stages as shown in Figure 5-9. The uniformity of power division from any input to any output can be within 3 dB, even for large numbers of inputs and outputs. The excess loss can be below 3 dB, again, even with large numbers of inputs

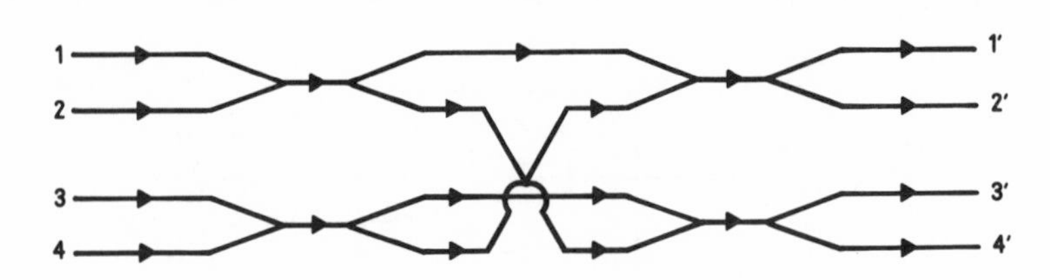

Figure 5-9 Multistage Star Coupler.

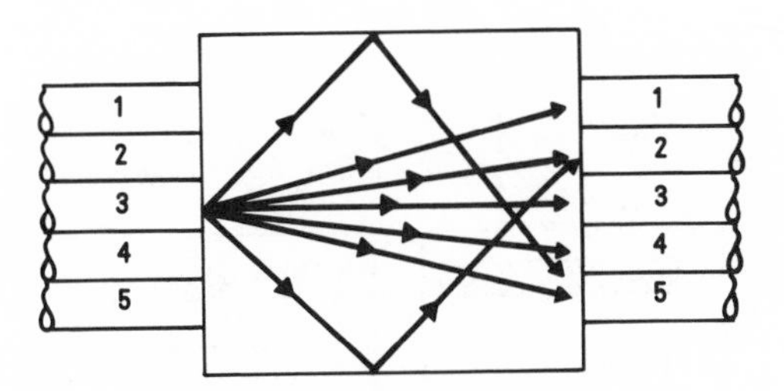

Figure 5-10 Slab Star Coupler.

and outputs. As with access couplers, fused tapered star couplers are mode sensitive, and therefore the power coupled between any given input and output port pair depends upon the distribution of the incoming optical power amongst the modes. Figure 5-10 shows a star coupler fabricated in the surface of a substrate by ion exchange. Light from the incoming fibers mixes in the common core of the slab waveguide. If the slab is sufficiently long, power division can be very uniform. Excess loss arises from absorption and scattering in the slab, and the fraction of the light which does not fall on the cores of the outgoing fibers. Figure 5-11 shows an integrated optics star coupler which is similar to the wavelength multiplexer described in 5.1.2 above. Here the coupled waveguides must be controlled to tight tolerances to obtain the desired coupling ratios. As with the wavelength multiplexer, this is a single mode device.

5.1.4 Mechanical Optical Switches

It is possible to fabricate an optical switch by mechanically moving parts within a coupler. Figure 5-12 shows an example, where a movable incoming fiber is coupled to one of two fixed outgoing fibers by mechanically moving a substrate holding the incoming fiber (between two stops). Figure 5-13 shows a similar design where a $1 \times N$ array of movable fibers is coupled to one of two $1 \times N$ arrays of fixed fibers, thus forming an N-pole double throw switch. Figure 5-14 shows a photograph of a

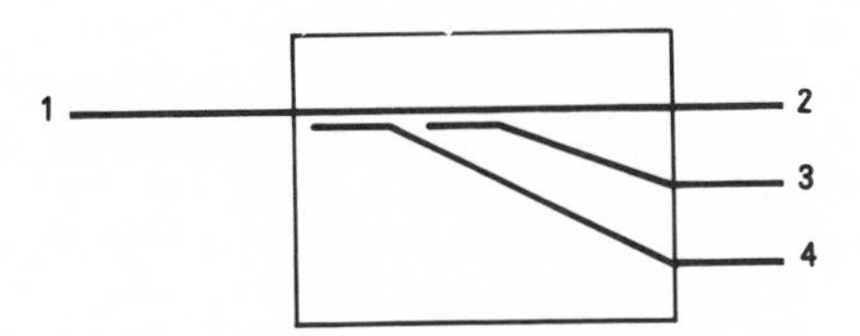

Figure 5-11 Integrated Optics Star Coupler.

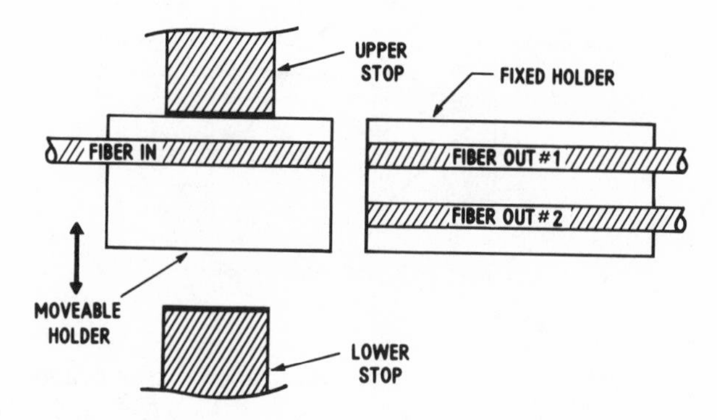

Figure 5-12 Mechanical Optical Switch Schematic.

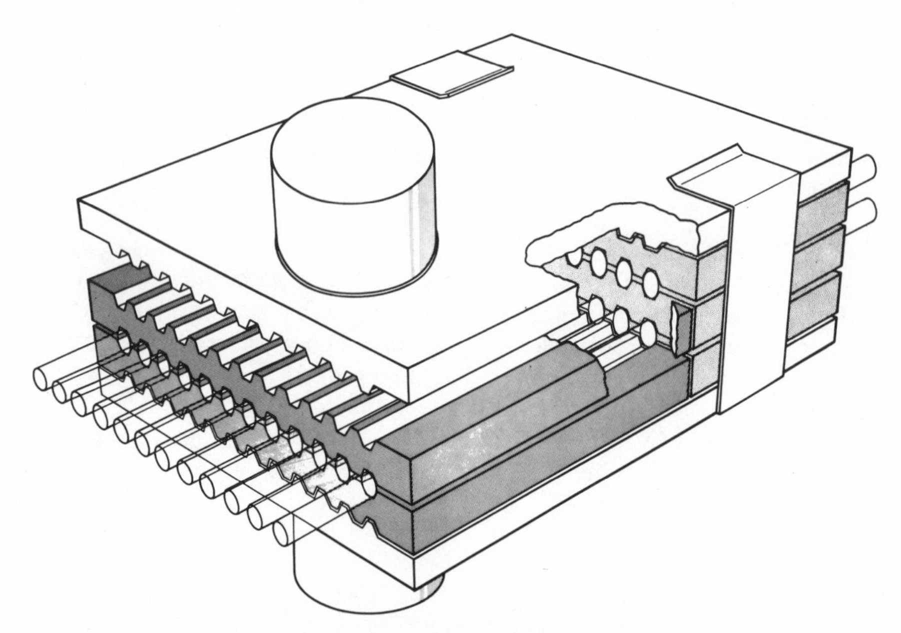

Figure 5-13 N-Pole Double Throw Mechanical Optical Switch. Courtesy of W. Young, Bell Communications Research.

Figure 5-14 Photograph of a 2 Stage 1 × 4 Mechanical Optical Switch. Courtesy of W. Young, Bell Communications Research.

two-stage 1 × 4 version of such a device. Mechanical switches of this type have been fabricated both from single mode and multimode fibers. They have been made with very low insertion loss (a small fraction of a decibel) and repeatably low loss over thousands of switching cycles. A device of this type is being considered as an option for redundancy protection in undersea fiber cable systems. Other mechanical switching devices have been fabricated using lenses to collimate the light beams and using movable mirrors or prisms to obtain switching action.

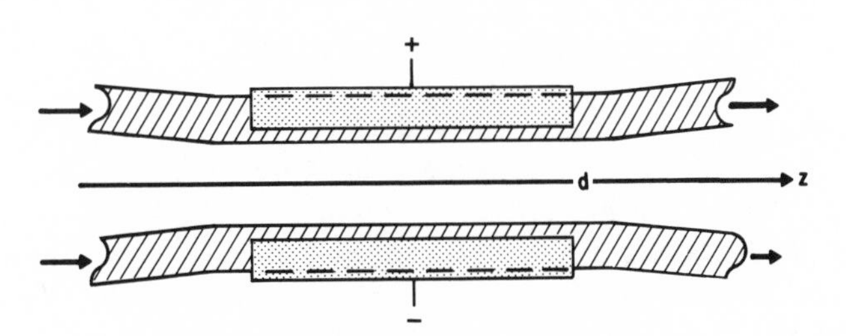

Figure 5-15 Integrated Optics Switch/Modulator.

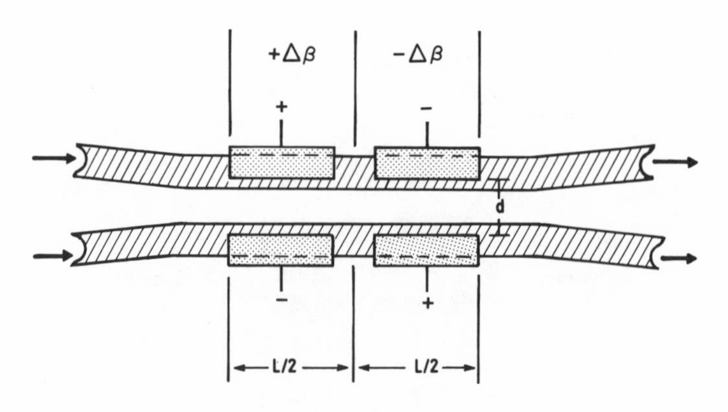

Figure 5-16 Δβ-Reversal Switch/Modulator.

5.2 Active Devices

One can fabricate active devices such as modulators and switches using materials whose optical properties (dielectric constant) can be modified by the application of an electric field or pressure from an acoustic field. Figure 5-15 shows an example of such a device, fabricated using integrated optics technology. Two coupled waveguides are formed in the surface of a material which is electro-optically active (e.g., lithium niobate). This material has a dielectric constant (tensor) which can be modified by the application of an electric field. The electric field is applied within the surface of the device by electrodes fabricated on the surface. The application of the electric field causes the coupling between the waveguides to change. Thus the device can act as a modulator (or switch) of the light passing from either input port to the two output ports. If the two waveguides are exactly alike, then one can obtain complete coupling from one to the other, no coupling, or anything in between. If the two waveguides are not identical (slightly different widths for example) then complete coupling from one to the other is not possible with this simple structure. However, Figure 5-16 shows a structure with two sets of electrodes which can obtain complete coupling even with nonidentical waveguides. The two sets of electrodes allow for two different applied voltages. This provides enough degrees of freedom to accommodate the mismatch of the coupled waveguides. Switches and modulators similar to those just described have been operated at very high modulation (switching) speeds (beyond 10 GHz) and with moderate modulating voltages (below 10 V). As with other integrated optics devices, they are presently in a research phase of emergence because of remaining problems with tolerance control, coupling efficiency to single mode fibers, polarization sen-

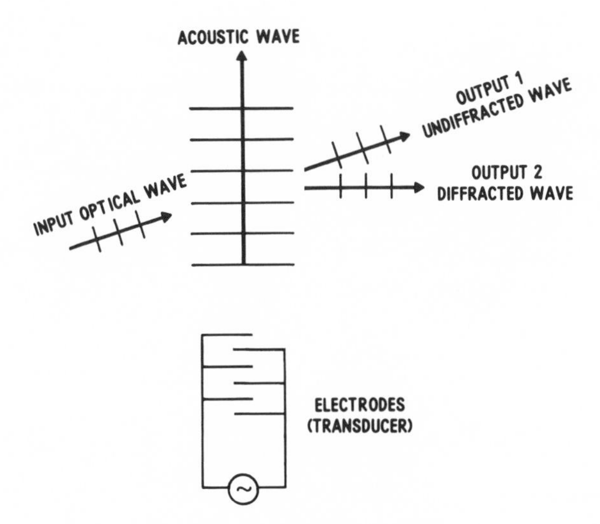

Figure 5-17 Acousto-optic Switch.

sitivity, and material stability. These devices work only with single mode optical signals.

Figure 5-17 shows an example of a switch which works on acousto-optic effects. An interdigitated transducer is used to launch a pressure wave into a piezoelectric material. This pressure wave travels across the substrate, and forms a moving grating in the surface of the substrate. That is, the acoustic pressure peaks and nulls modify the refractive index of the substrate surface. An incident optical beam is deflected (redirected) as it passes through this substrate by a mechanism called Bragg diffraction. The optical beam can be switched between two output ports by varying the frequency of the acoustic signal.

Problems

1. Consider a two-input two-output optical coupler as shown in text Figure 5-3. Assume that it has no excess loss, and that it satifies reciprocity. If the coupler is illuminated at port 1, then 10% of the light is emitted from port 3. If the coupler is illuminated at port 4, what fraction of light will be emitted from port 2?

   Define the transit loss of a coupler as the loss in decibels of light going from port 1 to port 2. What is the transit loss of this coupler? Define the access loss as the fraction of light entering from port 4, which is coupled to port 2 (in decibels). What is the access loss of this coupler?

2. Extending the ideas in Problem 1 above, let the access loss of a coupler be one of the following values: 20 dB, 17 dB, 13 dB, 10 dB, 6 dB, 3 dB.

   What are the corresponding transit losses for each of these access losses? (Assume no excess loss.)

6

System Phenomenology

There are a number of important effects (mostly harmful) which are apparent when optical components are assembled into a system. Several of these were not anticipated until they appeared in experimental or fully deployed systems, but could be explained by subsequent analysis.

6.1 Modal Noise

Modal noise is a phenomenon which occurs in multimode systems. It occurs when mode selective components in the system change their transmission (or other) characteristics in response to a randomly varying field pattern. Figure 6-1a shows a very simple example of the phenomenon. Two fibers a joined by an imperfect splice. A simple model of the loss of the splice would predict a 3-dB loss because the fiber cores overlap by 50% (in this two-dimensional example). Figure 6-1b shows two hypothetical field patterns (modes) which can propagate in the fiber cores. One field pattern is constant across the core. The other field pattern has a 180° phase shift (changes polarity) at the center of the core. The two field patterns are orthogonal. That is

$$\int_0^a E_1(x)\, E_2^*(x)\, dx = 0 \qquad (6\text{-}1)$$

If these field patterns are excited by the same optical source then they can add coherently to form a variety of other field patterns. Figure 6-1c shows two such linear combinations. Returning to Figure 6-1a, if combination A is incident upon the imperfect splice, then the loss through the splice will be infinite. If combination B is incident upon the splice, then the loss will be zero dB. Thus the loss of the splice depends upon the relative phase

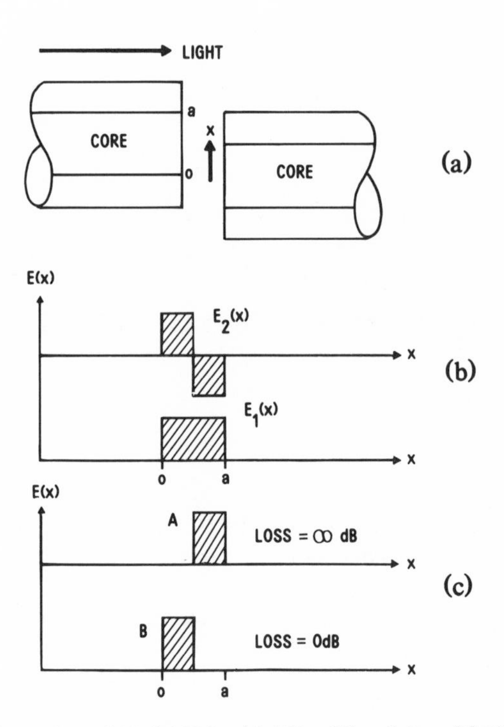

Figure 6-1 Illustrations of Modal Noise (a) Offset Fiber Splice; (b) Hypothetical Guided Mode Field Patterns; (c) Two Linear Combinations of the Modes of (b).

relationship of the two hypothetical modes as they arrive at the splice interface.

Suppose an optical source excites both modes at the input of the incoming fiber. Suppose the incoming fiber is 100 m long. This corresponds to 10^8 wavelengths at 1 μm wavelength. A slight change in the optical wavelength of the source can easily change the phase relationship of the two hypothetical modes by many radians as they propagate through this distance. Thus, as the wavelength emitted by the source randomly fluctuates (slightly) or as the fiber undergoes slight mechanical disturbances, the loss at the splice will fluctuate widely.

The above is a simple example of the fluctuations which can occur in the loss of a mode sensitive (imperfect) splice or the access loss of a mode selective access coupler due to fluctuations in the relative phases of the propagating modes which make up the arriving light. These fluctuations appear in systems using laser sources, but do not appear in systems using LED sources. The reason for this is that the LED emits light signals in each of its spatial output modes which independently randomize their amplitudes and phases very rapidly (the correlation time is less than picoseconds). As a result of all of this random fluctuation (modal noise) is averaged out over any time interval of practical interest. Systems using lasers and operating at high bandwidths (e.g., high digital data rates) do not benefit from this averaging effect. In some system applications, using lasers, attempts are made to induce averaging, in short time intervals, by superimposing high frequency modulation on the ordinary information bearing modulation. This approach is based on the hope that the high frequency superimposed modulation will cause the laser wavelength to vary rapidly, in turn averaging out the modal noise effects. It is a useful technique, but not totally effective.

Modal noise can also be caused by variations in the responsivity across the surface of a detector. These variations are typically small, and the effect is generally negligible in digital systems. However, for analog modulation applications where very high signal-to-noise ratios are often required, even this source of modal noise can be a serious problem.

As mentioned above, modal noise occurs in multimode systems. However, even a single mode fiber has two polarizations which are, in effect, two separate modes. If the fiber is birefringent (polarization sensitive) and if there are polarization selective components in the system (other than the fiber) then it is possible to experience modal noise.

In addition, single mode fibers will propagate lossy higher order modes for short distances. If a single mode fiber is placed between two closely spaced splices, then it is possible for interference between the low loss mode and a lossy higher order mode to cause modal noise. (The higher order mode is excited at the first imperfect splice and interferes at the second imperfect splice.)

Modal noise is caused primarily by imperfect splices and can occur even in relatively short end-to-end links. Modal noise is increasingly important at higher data rates (in digital systems) because there is less time for averaging of the noise to occur within a bit interval. Thus, even on short links with adequate bandwidth, increasing the bit rate can result in the emergence of modal noise problems.

6.2 Mode Partition Noise

In Section 3.1.1 above we described the spectral characteristics of laser diodes, and, in particular, the fact that the laser could distribute its output power among a number of optical frequencies (longitudinal modes) in an unpredictable manner. In Section 2.1.4 above we described the material dispersion of fibers, where different wavelengths travel through the

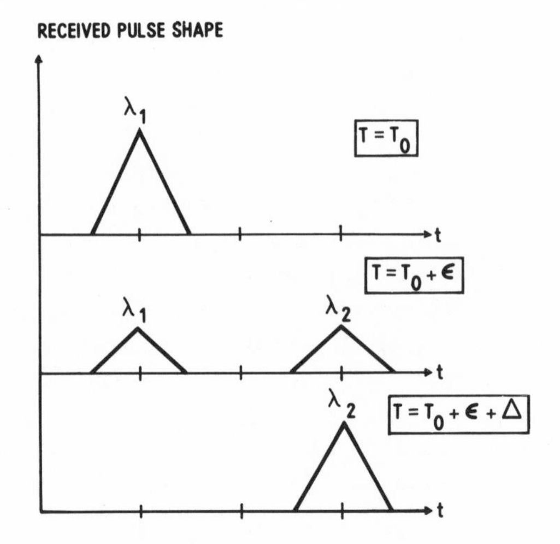

Figure 6-2 Illustration of Mode Partition Noise-Received Pulse Shapes vs Time.

fiber at different group velocities. These two effects can combine to produce a phenomenon called mode partition noise. Figure 6-2 shows an example of this effect. A laser is assumed, in the figure, to emit light in three different ways, at three different times. Initially the laser is shown emitting light at a single frequency (with some small amount of power in some longer wavelength longitudinal modes). This light is modulated by a triangular pulse as shown, and arrives at the output of a fiber as a similar triangular shaped pulse. At a later time, the laser is emitting simultaneously in two longitudinal modes. The modulation of the laser is again triangular shaped, but because the two wavelengths travel at different group velocities, the fiber output pulse is a sum of two separated triangular waveforms. At a still later time, the laser is oscillating in a single longitudinal mode again, but at a longer wavelength than it was initially. The modulation is again triangular, but the fiber output pulse is delayed from what it was initially because of the slower group velocity of the longer wavelength. We see then that as the laser redistributes its power unpredictably among several longitudinal modes (wavelengths), the pulse produced at the fiber output will change shape and move in time relative to its nominal position in the center of a time slot. This noiselike disturbance of the shape and position of the received pulse can cause errors in digital systems (if the movement is a substantial fraction of a time slot) and noise in analog systems.

For example, consider a digital system operating at a wavelength of 0.82 μm at a data rate of 300 Mb/s. At this wavelength, the fiber material dispersion is 100 picoseconds of delay change per kilometer of transmission per nanometer of wavelength change. The width of a time slot is 3.333 ns. Suppose we wish to keep the arrival time fluctuations due to mode partition noise below 20% of a time slot (The actual allowable amount depends upon how much performance degradation one is willing to allocate to mode partition noise, and upon the details of the laser longitudinal mode hopping behavior.) This corresponds to 0.666 ns of allowable delay difference. Suppose the laser output is randomly distributed among 9 longitudinal modes separated by 0.5 nm in wavelength. The wavelength change between the shortest and longest mode is then 4 nm. With the material dispersion given by 100 ps/km-nm; this corresponds to a delay spread between the fastest and slowest wavelength of 400 ps (0.400 ns) per kilometer. Thus we see that in just a little over 1.5 km of transmission the time jitter of mode partition noise will exceed our 0.666-ns allowance. Therefore, with these assumptions, the transmission distance would be limited to 1.5 km, regardless of the signal power level after this distance of transmission. The above example is for illustration. Longer distances of transmission can be achieved by operating at a wavelength where the material dispersion is smaller (near 1.3 μm), or by using a laser with a narrower longitudinal mode spectrum.

Recently, experiments have been performed to determine how small a satellite longitudinal mode must be in order to neglect it relative to dominant longitudinal modes. Theory and experiment have shown that a small satellite mode may contain a substantial fraction of the laser output for a small percentage of the time. Thus if one is trying to achieve a low bit error rate in a digital system one must be concerned that errors may be caused during the infrequent periods when these apparently weak satellite modes contain relatively large amounts of power (compared to their average power). Based on experiments, rules of thumb have emerged regarding the safe level below which one can neglect a satellite mode even if one is trying to achieve a low bit error rate (10^{-10}). Whether or not these rules of thumb will be changed as more experimental data is accumulated, and whether or not these rules of thumb apply to all lasers, remains to be seen. The present rule of thumb is that a satellite mode must have an average power level of less than a hundredth of the largest longitudinal mode average power level in order to be neglected from mode partition noise considerations.

6.3 Modal Distortion

Modal distortion is a phenomenon which is closely related to mode partition noise. In our discussion of mode partition noise in Section 6.2

above we assumed that the laser would redistribute its energy among a number of longitudinal modes in a random manner. It is possible that as a laser is directly modulated (the current is modulated) the laser output may jump between two or more adjacent longitudinal modes (wavelengths) synchronously with the modulation. This can cause amplitude modulation to be converted to delay modulation at the end of a fiber with material dispersion. The effect manifests itself as a nonlinearity in the end-to-end link, which can be important in analog systems.

For example consider the following situation. A laser can oscillate in two longitudinal modes which are separated by 0.5 nm in wavelength. The laser light propagates through a fiber with 100 ps/km-nm of dispersion at the nominal wavelength and which is 1 km long. Thus the delay difference for the two laser longitudinal modes is 50 ps. The laser is modulated sinusoidally about its average output level at a frequency of 100 MHz. As the laser is modulated its output power shifts between the two longitudinal modes synchronously with the modulation. Thus the total fiber input power and the fiber input powers in the two longitudinal modes (1 and 2) are given by

$$p_{\text{Total}}(t) = P_o[1 + \cos(200\pi t)]$$

$$p_1(t) = \frac{1}{2} P_o[1 + \cos(200\pi t)] \cdot [1 + \cos(200\pi t)] \qquad (6\text{-}2)$$

$$p_2(t) = \frac{1}{2} P_o[1 + \cos(200\pi t)] \cdot [1 - \cos(200\pi t)]$$

where P_o is the average total power, and t is in microseconds.

Note that this assumption is for illustration, and does not represent the detailed behavior of an actual laser. The total power at the output of the fiber (neglecting loss) is given by

$$p_{\text{Total output}} = \frac{1}{2}\left\{\begin{matrix} P_o[1 + \cos(200\pi t)] \cdot [1 + \cos(200\pi t)] + \\ P_o[1 + \cos(200\pi t')] \cdot [1 - \cos(200\pi t')] \end{matrix}\right\}$$

where

$$t' = t - 5 \times 10^{-5} \text{ (microseconds)} \qquad (6\text{-}3)$$

With a little algebra we can rewrite this as follows:

$$P_{\text{Total output}} = P_o[1 + \cos(200\pi t) + 10^{-2}\,\pi \sin(200\pi t)\cos(200\pi t)]$$

$$(6\text{-}4)$$

Thus we see that the conversion of direct modulation of the laser power to delay modulation introduces a second harmonic at the receiver, whose amplitude (relative to the fundamental) is proportional to the product of the delay difference of the two longitudinal modes and the modulation frequency. In this example we obtain a second harmonic which is 30 dB below the fundamental; but other values would be obtained for different assumptions.

6.4 Laser Noise Caused by Reflections

The laser is a resonant cavity whose properties can be changed by the introduction of external mirrors. Reflections can be coupled back into the cavity from the surface of an attached fiber or from splices which are not perfectly index matched. With low loss fibers, reflections of substantial level can return from locations which are billions of wavelengths away from the laser cavity. These reflections arrive in unpredictable and changing phase relationships relative to launched light. Experiments and theory have shown that these reflections can enhance the intrinsic power fluctuations in the total laser output, and in the outputs of individual longitudinal modes, and can induce wavelength (longitudinal mode) jumps. Methods have been employed to reduce these reflections by reducing the coupling efficiency between the laser and the fiber, particularly when single mode fibers are used. (In a multimode fiber, reflections from distant locations are likely to be distributed amongst all of the fiber modes, and thus will not couple strongly with the single mode laser cavity.)

PART II: APPLICATIONS

Having reviewed the technology in part 1 of this book we are now ready to discuss the applications, both present and projected, and to understand how the capabilities and limitations of the system components translate into capabilities and limitations of the systems which are assembled from those components.

7

Telecommunications Trunking

When research began in modern optical fiber technology in the late 1960s the most important application envisioned was in the replacement of metallic cables (containing twisted wire pairs) which carry telephone circuits between telephone buildings. Figure 7-1 is a simplified diagram of a telephone network. The wires which run from a customer's premises to a telephone building are called loops. The circuits which go between telephone buildings (over wires or other media) are called trunks. Within the telephone building requests for service are answered by using a switching machine to connect the customer's loop to the required trunk (to carry the call to its destination). Thus, while loops are dedicated to the customer (except for party lines), trunks are shared by all customers and allocated when needed.

In the early days of telephony voice signals were transmitted between central offices over individual pairs of wires of appropriate gauge (to provide an acceptable loss). With the advent of modern electronics, it became economical to transmit several voice circuits simultaneously over a single pair of wires using a carrier system. A carrier system is shown in Figure

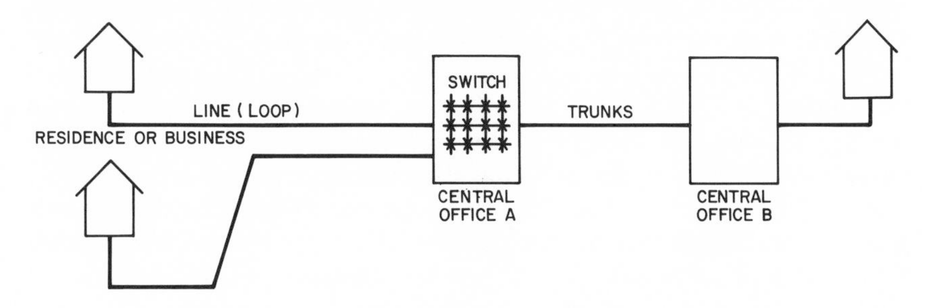

Figure 7-1 Simplified Diagram of a Telephone Network.

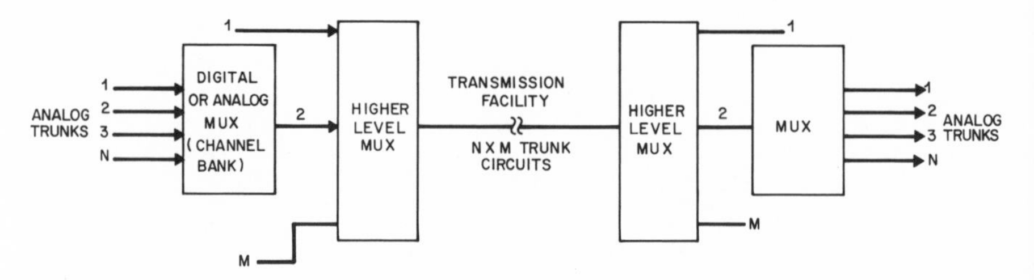

Figure 7-2 Carrier System.

7-2. Terminals at either end of the system present wire pairs to the switching machines.

Individual voice circuits are combined within the terminals using frequency division or time division multiplexing techniques, to produce a composite electrical signal which is carried between terminals by wire pairs, coaxial cable, a microwave radio link, or fibers. Individual voice circuits are allocated a bandwidth of 4 kHz. If the circuits are combined using frequency division multiplexing, then they are simply stacked up in a band of frequencies of approximately 4 × N kHz width (where N is the number of circuits). For example a "group" of 12 circuits occupies approximately 48 kHz of bandwidth (using single sideband stacking). If the circuits are combined using time division multiplexing, then each circuit is allocated 64 kb/s of digital rate (8000 samples per second and eight bits per sample). Thus a "digroup" of 24 circuits requires about 1.5 Mb/s of total digital rate. The communication system which includes the terminals and the transmission medium is called a transmission facility. Each voice circuit carried over the transmission facility is called a trunk circuit. A collection of trunk circuits between two locations is called a trunk group.

7.1 T-Carrier

In metropolitan areas, transmission facilities are frequently implemented using a very popular approach called T-carrier. T-carrier was first introduced in 1962 as a means for provision of trunk circuits using digital transmission over wire pairs. T-carrier was designed to be compatible with the existing wire pair cables installed in ducts under the streets, and with the existing locations of underground "manholes" placed approximately every 6000 ft (2 km). Based on measurements of the existing wire pair plant, it was determined that a digital rate of 1.5 Mb/s could be supported with a 6000 ft nominal spacing between digital repeaters. Figure 7-3 shows a diagram of a typical T-carrier system. The terminals which convert 24 individual analog voice circuits to digital form and which combine

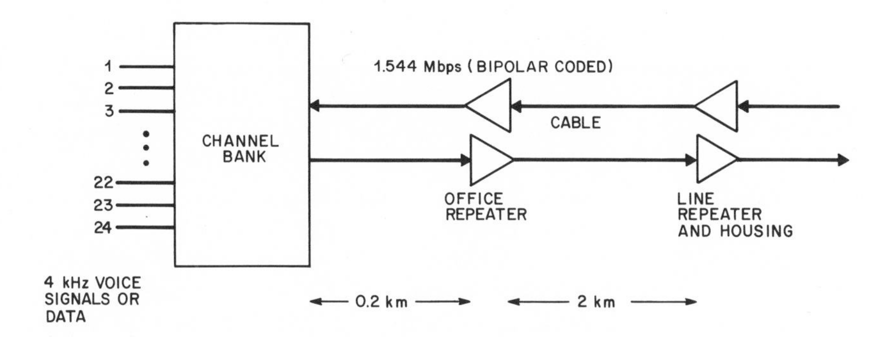

Figure 7-3 Typical *T* Carrier System.

these into a composite 1.544-Mb/s digital stream are called channel banks. The 1.544-Mb/s digital output stream is called a DS1 signal (digital signal 1). Repeaters are housed in environmentally sealed boxes, called apparatus cases (or repeater housings), which are located in unattended manholes spaced at approximately 6000 ft. T-carrier is a very reliable technology which has been cost-reduced several times. Figure 7-4 shows a photograph of a two-way T-carrier repeater. It can be purchased in quantity for less than 50 dollars.

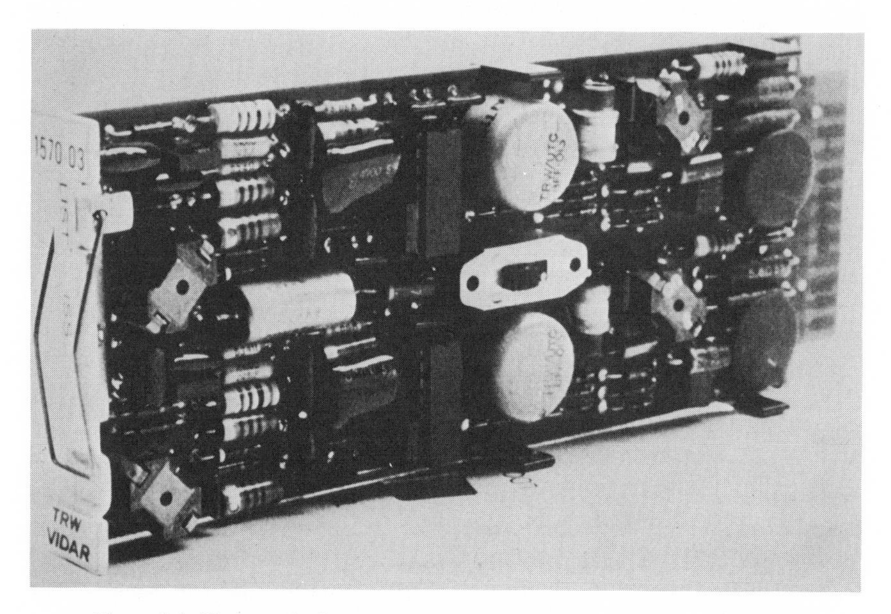

Figure 7-4 Photograph of a Typical *T*-Carrier Repeater. Courtesy of TRW Inc.

In spite of its popularity, T-carrier has certain shortcomings. In crowded metropolitan areas, existing duct space (for wire cables) is exhausted and manholes are crowded with repeater housings and cables. The cost of installing new ducts is prohibitive, and the disruptions associated with such construction are often not tolerable. Attempts to design advanced versions of T-carrier with more channels per wire pair have met with limited success, because the attenuation of the wires increases rapidly with frequency (due to the skin effect). Furthermore, crosstalk (interference) between DS1 signals in the same cable and in the same repeater housing increases with frequency to intolerable levels. Replacement of the existing cables with cables containing larger gauge wires offers limited potential, because the total bit-rate capacity of a wire cable is related to its cross sectional area, independent of the type of wire it contains (including coaxial tubes). The proposal of C. Kao and co-workers that glass fibers could carry hundreds of telephone conversations for miles without repeaters held out the promise for relieving these problems. In addition to duct congestion, T-carrier has other shortcomings. The repeaters are reliable, but, when they do fail, the location of a failed repeater must be determined, and craft must be dispatched to effect repairs. Repairs require manholes to be entered and sealed apparatus cases to be opened. This can be a time-consuming and expensive process. T-carrier is also used in suburban and rural areas where the cables are often mounted on poles, and repeater housings are often pole-mounted as well. In these situations, the housings are subject to vandalism, and lightning striking cables can damage repeaters (even though protection devices are employed). Fiber optic trunk transmission facilities can be designed to be insensitive to lightning and can have fewer unattended repeaters and housings (sometimes none) to maintain.

7.2 Fiber Optic Trunk Transmission Facilities [9, 10, 11, 24]

Figure 7-5 shows a diagram of a typical fiber optic trunk transmission facility. Conventional channel banks interface with individual analog voice circuits and convert these, in digroups of 24, to DS1 signals (1.544 Mb/s). Multiplexers combine several (or many) DS1 signals to produce a composite digital signal at a higher clock rate (e.g., 44.7 Mb/s, 135 Mb/s, etc.) so that the bit rate capability of the fiber can be economically shared among more than 24 trunk circuits (assuming more than 24 trunk circuits are required). The composite high-clock-rate digital signal is carried over the fiber medium.

When designing or choosing a fiber optic trunk transmission facility one is faced with a number of choices for the optical components (fiber,

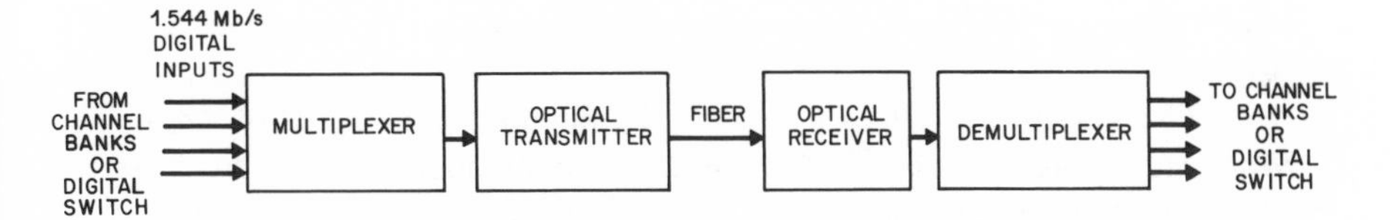

Figure 7-5 Typical Fiber Optic Trunk Carrier Facility.

source, detector, etc.). One can choose a multimode fiber or a single mode fiber; short or long wavelength sources and detectors; LED or laser sources; and *p-i-n* or APD detectors. One must also select the modulation rate for each optical signal, and one can consider such things as wavelength multiplexing. An exhaustive list of combinations would contain many of little interest. Figure 7-6 contains a tabulation of six important system choices in order of increasing performance and increasing technical sophistication. (We list the newer long wavelength technology as more sophisticated than short wavelength technology, but one could argue as to whether a long wavelength LED is more sophisticated than a short wavelength laser.) In the discussion to follow next we will derive the performance, in terms of achievable transmission distance as a function of data rate, for each of these system options. We shall take the longest amount of text to derive the performance of the first system. Then we will build on that to derive the performance of the others, concentrating on the things that change.

7.2.1 Short Wavelength/Multimode/LED/*p-i-n* Systems

The least technically sophisticated digital fiber optic transmission system would employ an LED source, a multimode fiber, and a *p-i-n* photodiode detector. A system using the short wavelength (0.8–0.9 μm) region

SOURCE	WAVELENGTH	FIBER	DETECTOR	DATA RATE	REPEATERLESS SPAN
LED	SHORT	GRADED	PIN	MODERATE	SHORT
LASER	SHORT	GRADED	APD	HIGH	MODERATE
LED	LONG	GRADED	PIN	MODERATE	MODERATE
LASER	LONG	GRADED	PIN	HIGH	LONG
LASER	LONG	SINGLE MODE	PIN	HIGHER	LONGER
LASER	LONG	SINGLE MODE	APD	HIGHER	LONGEST

Figure 7-6 Tabulation of Fiber Trunk Carrier Facility Alternatives.

would draw on the more mature GaAlAs source and Si detector technology.

The GaAlAs LED can be directly modulated at rates of up to 50 Mb/s (for typical devices). The peak power launched by a typical GaAs LED into a multimode fiber with a 50 μm core and a 0.2 NA is approximately 50 μW. A GaAlAs LED emits light over a band of optical wavelengths having a full width to half-maximum of about 50 nm, resulting in pulse spreading due to material dispersion of about 5 ns/km. (The material dispersion in this wavelength region is around 100 ps/nm-km.) The broad spectral width of the LED corresponds to rapidly fluctuating amplitudes and phases of the emitted spatial modes, causing modal noise effects to average to negligible levels. The output power of the LED decreases with temperature as shown in Figure 3-4, but this is generally accommodated by the dynamic range capabilities of the optical receiver. That is, local measurements of the LED output, accompanied by closed loop control, are not generally employed. LEDs will operate at elevated temperatures with reduced power and reduced lifetimes, but controlled ambients are not required in most applications.

The loss of a cabled fiber in the 0.8—0.9 μm band of wavelengths is typically about 3—6 dB/km, depending upon the grade (price) and the temperature range of operation (microbending). A standard graded index fiber with a 50 μm core and a 0.2 N.A. (1% index difference between the axis of the fiber and the edge of the core) will have a modal delay spread of 0.5—5 ns/km depending upon the grade of the fiber (difference between the delay of the slowest and the fastest modes). Thus the pulse spreading due to material dispersion and the broad spectral width of the LED (5 ns/km full width to half-maximum) will typically dominate the modal delay spread. Recall that the total pulse spreading is the square root of the sum of the squares of the two effects. Thus if the material dispersion effect is 5 ns/km and the modal delay spread effect is 2 ns/km, then the total pulse spreading due to propagation through the fiber will be 5.39 ns/km. The fiber is not sensitive to electromagnetic interference. It does not pick up light from the side, and it does not radiate significant amounts of light unless it is bent in a small radius. Heavy doses of radiation can cause the fiber loss to increase, but this is a concern only in special environments.

The silicon *p-i-n* detector will respond faithfully to light modulated at rates well beyond 100 MHz (typically). A receiver constructed with a silicon *p-i-n* detector and a low noise preamplifier will typically require an optical signal level of about 5000—40,000 photons per received pulse to achieve a 10^{-9} bit error rate. Since the energy in a photon at this wavelength is about 2×10^{-19} J, the average power required at the receiver for a bit rate of B (bits) is given by

$$P_{\text{av required}} = \frac{1}{2} N_r \times 2 \times 10^{-19} \times B \text{ (watts)} \qquad (7\text{-}1)$$

where N_r is the required number of photons per received pulse. The factor of 1/2 is based on the assumption that the received optical signal contains, on the average, half "ones" and half "zeros."

Since the silicon detector has a large volume in which carriers are produced, high levels of radiation can cause undesired dark current. However, since a receiver employing a silicon *p-i-n* detector is limited by preamplifier noise, rather than shot noise from dark current, this is only a concern in special applications.

We can now consider a system example. Figure 7-7 shows a listing of the assumed parameters. We assume that the average output of the GaAs LED into a standard graded index multimode fiber is −15 dBm (31 μW), assuming that the LED is turned on half of the time. (That is we assume half "ones" and half "zeros.") We assume that the energy required at the receiver is 20,000 photons per received pulse, to achieve a 10^{-9} error rate. Thus the average receiver power required is $-117 + 10 \log(B)$ (dBm), where B is the data rate in bits per second.

Knowing the average transmitter output power and the required receiver input power (given B), we can subtract these two (in dB) to obtain the allowable loss between the transmitter and the receiver. This loss must be allocated to cabled fiber, splices, connectors, and margin.

We assume that the cabled and spliced fiber has a loss of 5 dB/km at the temperature of operation where the loss is highest. We assume that

LED OUTPUT: −15 dBm AVERAGE* POWER INTO A 50 μm CORE, 0.2 NA FIBER

RECEIVER SENSITIVITY: 20,000 PHOTONS/BIT ⇒ $1/2 \times 2 \times 10^{-19} \times 20{,}000 \times B$ (WATTS)
$= 2 \times 10^{-15} B$ (WATTS) $= -117 + 10 \text{ LOG } (B)$ dBm

MARGIN = 10 dB OR 3 dB

FIBER + SPLICE LOSS = 5 dB/km

CONNECTOR LOSSES = 6 dB TOTAL

PULSE SPREADING = 5 nsec/km (FWHM) ⇒ MAX BIT RATE $= 0.5/5 \times 10^{-9} L$ (Hz)
$= 1.0 \times 10^{8}/L$ (Hz)

*50 % DUTY CYCLE

Figure 7-7 Sample System Parameters — Short Wavelength LED System with *p-i-n* Detector and Multimode Fiber.

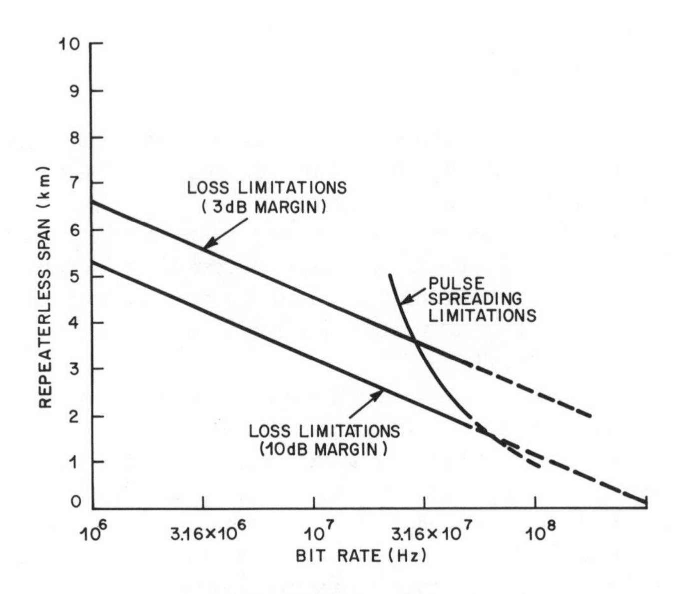

Figure 7-8 Repeaterless Span vs Data Rate — Short Wavelength LED System with *p-i-n* Detector and Multimode Fiber.

the total connector loss at both ends of the link is 6 dB. The choice of the value of margin in the link loss budget deserves some discussion. The purpose of margin is to accommodate degradations in time which are anticipated, so as to extend the mean time between required maintenance actions. However, there is no point in allocating excessive margin, since circumstances which would necessitate that large margin would imply imminent system failure. We might anticipate that the LED output could fall by 1.5—3 dB below its nominal value, as a result of aging and temperature effects. We might anticipate a small variance of the receiver performance (say 1—2 dB) from its nominal value. We might allow a few decibels for additional cable splices needed to implement repairs. In this example we shall pick two values of margin and show two sets of calculated results. The two values are 3 dB (optimistic) and 10 dB (conservative).

Using these results we can calculate the allowable distance of transmission (maximum spliced cable loss) as a function of the data rate. Curves for the two values of margin are shown in Figure 7-8. We observe that the allowable transmission distance decreases by 0.6 km each time the data rate is doubled. This is because the required average power at the receiver increases by 3 dB each time the rate is doubled, and the loss of the spliced cabled fiber is 5 dB/km. The curves are shown dotted beyond 50 Mb/s because it is difficult to modulate the GaAs LED at these higher rates. Note that the effect of the 7 dB difference in the two choices for system margin corresponds to 1.4 km of allowable transmission distance.

We observe that, even with this technology, the allowable distance of transmission (without a repeater) is 4.8–6.2 km at 1.5 Mb/s, 2.4–3.1 times the distance achievable with T-carrier on wire pairs. At 25 Mb/s the allowable transmission distance without repeaters is 2.4–3.8 km.

The allowable distance of transmission can be limited by pulse spreading as well as loss. When pulse spreading causes intersymbol interference (interference between adjacent pulses) the receiver performance deteriorates rapidly. In wire pair systems, intersymbol interference can be compensated by using equalizers, which enhance the high frequency content of the received pulses (thus making them narrower again). In optical fiber systems equalization is of little or no help, because the process of enhancing high frequencies in the received pulses also rapidly enhances high frequency noises and rapidly uses up the link loss budget. One can analyze the reduction in receiver sensitivity as a function of intersymbol interference in great detail. The results of such calculations lead to the following rule of thumb. To keep the receiver sensitivity penalty below 1 dB, the full width to half-maximum of the pulse spreading caused by the fiber should be below 50% of the spacing between pulses. In this example, we assume that the pulse spreading is 5 ns/km. Thus, we obtain the following limitation on the distance of transmission (L [km]):

$$5 \times 10^{-9} LB \leqslant 0.5 \tag{7-2}$$

where B is the data rate in bits per second ($1/B$ is the pulse spacing). This limitation is shown in Figure 7-8 as the curve labeled pulse spreading limitations. We see that this limitation causes the allowable repeater spacing to be cut in half every time the data rate is doubled. However, for the assumptions chosen in this example, the allowable transmission distance (without repeaters) is limited by loss for data rates up to 30 Mb/s, near the practical modulation limits of GaAs LEDs. If a lower loss were selected for the cabled/spliced fiber and/or less link loss budget were allocated to connectors and margin, then pulse spreading might become a limitation at the higher practical data rates.

7.2.2 Short Wavelength/Multimode/Laser/APD Systems

In order to achieve improved performance (higher data rates and/or longer distances of transmission) one can use a GaAlAs laser and a silicon APD is a substitute for the LED and *p-i-n* devices discussed in Section 7.2.1 above. One could also use the laser with a *p-i-n* detector, but the complexity and cost of the laser usually warrants the use of the more sensitive APD detector. The use of an LED with an APD detector is another possible choice.

The laser will launch approximately 0 dBm of average optical power into a standard graded index optical fiber. This is about 15 dB more than what can be achieved with a typical GaAs LED.

The laser can be modulated at rates beyond 1 Gb/s with careful design of the transmitter electronics.

The laser requires a more complicated driver circuit and is more sensitive to electrical abuse and to temperature, as described in Sections 3.1.1 and 3.3.2 above.

Systems employing laser sources are also susceptible to modal noise and mode partition noise as described in Sections 6.1 and 6.2 above. We shall ignore these potential noise problems in the calculations below, but they can be important when actual systems are implemented, depending upon the laser characteristics.

A receiver employing a silicon APD is typically about 10–15 dB more sensitive than a receiver employing a *p-i-n* detector, because of the ability of the avalanche gain process to partially overcome the limitations of preamplifier noise. The APD receiver also has the potential for increased dynamic range because the APD gain is adjustable (over a limited range), and because the availability of avalanche gain allows the preamplifier designer more freedom in trading noise performance against dynamic range performance.

On the other hand, the APD requires a high voltage bias supply (usually implemented with a small dc–dc converter) which must be adjusted to track out the temperature dependence of the avalanche gain. Additionally, dark current caused by radiation (in high radiation environments) can reduce the sensitivity of the receiver by causing additional noise or by producing excessive current.

Figure 7-9 shows a set of typical parameter values for a system employing a GaAlAs laser and a silicon APD.

We assume that the average power coupled into the multimode fiber by the laser is 0 dBm (assuming half "ones," half "zeros," and a non-

LASER OUTPUT: 0dBm AVERAGE INTO 50μm CORE, 0.2 NA FIBER
RECEIVER SENSITIVITY: −130 + 10 LOG B (dBm)
MARGIN: 10 dB OR 3 dB
FIBER + SPLICE LOSS: 5 dB/km
CONNECTOR LOSSES: 6 dB TOTAL
PULSE SPREADING: 1.0 nsec/km (FWHM) ⇒ MAX BIT RATE = 5×10^8/L

Figure 7-9 Sample System Parameters — Short Wavelength Laser System with APD Detector and Multimode Fiber.

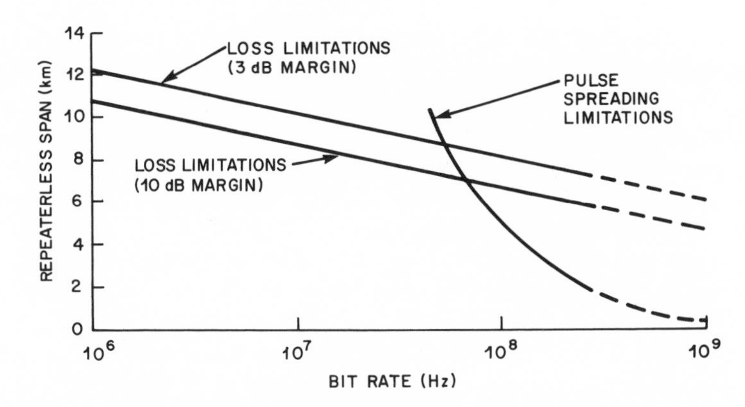

Figure 7-10 Repeaterless Span vs Data Rate — Short Wavelength Laser System with APD Detector and Multimode Fiber.

return-to-zero modulation format). (Note: in high data rate systems employing lasers, a return-to-zero modulation format is often used to suppress the tendency of the laser to oscillate if left in the "on" state for long periods of time, and also to combat modal noise by forcing multi-longitudinal mode behavior. This reduces the average laser output by 3 dB.)

We assume that the receiver incorporating the APD requires 1000 photons per received optical pulse to achieve a 10^{-9} bit error rate. This is 13 dB more sensitive than what we assumed in Section 7.2.1, in our system example where a *p-i-n* detector was used.

The required average power at the receiver is then given by -130 dBm $+ 10 \log (B)$; where B is the data rate in bits per second.

We assume the same values of cabled/spliced fiber loss as in the system example given in Section 7.2.1: 5 dB/km. Likewise we use the same value for connector loss: 6 dB total (both ends); and the same two values for margin: 3 or 10 dB.

We assume that since we are using a laser source, the pulse spreading is limited by modal delay spread (rather than dispersion), which we assume to be 1 ns/km. Since the dispersion of the fiber is 100 ps/km-nm at these wavelengths, this implies that the spread in the wavelengths of the significant laser longitudinal modes is below 5 nm (about 10 modes). (Consistent with our discussion in Section 6.2, we define insignificant modes as those with average power 20 dB below that of the dominant mode.)

Using these assumptions we can calculate the maximum distance of transmission (without repeaters) based on loss, and based on pulse spreading. These results are shown in Figure 7-10. We see that at a data rate of

25 Mb/s the allowable repeaterless span is 8–9.4 km. We see that at about 60 Mb/s pulse spreading becomes the limiting factor on repeaterless span (for the parameter values used here). The curves are shown dotted beyond 300 Mb/s because of concerns with the effects of modal noise and mode partition noise on the system performance.

7.2.3 Long Wavelength/Multimode/LED/*p-i-n* Systems

With the advent of long wavelength technology (InGaAsP sources and Ge or InGaAsP detectors) we can take advantage of the lower fiber loss and the material dispersion minimum at 1.3 μm wavelength.

The properties of the long wavelength LED and the long wavelength *p-i-n* detector are similar to their short wavelength counterparts (discussed in Section 7.2.1 above) with the following exceptions.

Long wavelength LEDs are fabricated with lattice matched layers, resulting in a reliability which has been estimated to exceed 10^9 hours of life (interpreted as the reciprocal of the failure rate) for properly made devices at room temperature. This is about three orders of magnitude more than estimated lifetimes of GaAlAs devices (which cannot be lattice matched between heterogeneous layers). The long wavelength LED is somewhat more temperature sensitive (compare Figure 3-4 to Figure 3-5), but this is not considered to be a significant problem. Although the device output decreases more rapidly with temperature than for the GaAlAs LEDs, the devices continue to operate at temperatures well above 200°C, and their long lifetimes accommodate the lifetime reduction associated with elevated temperature operation.

Long wavelength detectors have significantly more leakage current than silicon devices, but in receivers employing *p-i-n* detectors preamplifier noise usually dominates the shot noise of the leakage (dark) current. However, shot noise from leakage current can be a problem at elevated temperatures, depending upon the device and the data rate.

LED OUTPUT: −15 dBm AVERAGE POWER INTO A 50 μm CORE, 0.2 NA FIBER

RECEIVER SENSITIVITY: −119 + 10 LOG (B) dBm

MARGIN = 10 dB OR 3 dB

FIBER + SPLICE LOSS = 2 dB/km

CONNECTOR LOSSES = 6 dB TOTAL

PULSE SPREADING = 1.0 nsec/km (FWHM) ⇒ MAX BIT RATE = $5 \times 10^8/L$ (Hz)

Figure 7-11 Sample System Parameters—Long Wavelength LED System with *p-i-n* Detector and Multimode Fiber.

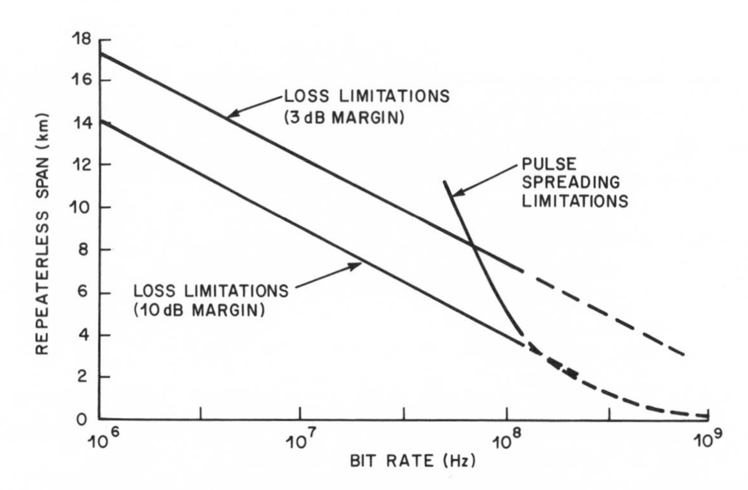

Figure 7-12 Repeaterless Span vs Data Rate — Long Wavelength LED system with *p-i-n* Detector and Multimode Fiber.

Long wavelength *p-i-n* detectors usually can be biased from voltages needed for conventional electronics, eliminating the dc—dc converter often required with silicon detectors.

Figure 7-11 gives a tabulation of typical parameter values for a long wavelength system employing these devices.

The LED is assumed to launch −15 dBm of average power into a standard multimode fiber. The receiver is assumed to require 20,000 photons per received pulse to achieve a 10^{-9} bit error rate. This translates into a required average optical power at the receiver of −119 dBm + 10 log (B), where B is the data rate in bits per second. Note that the photons are 2 dB less energetic at 1.3 μm wavelength than at 0.9 μm wavelength. The cabled fiber loss, including splices, is assumed to be 2 dB/km at the anticipated ambient temperature of highest loss. Two values of margin: 3 dB and 10 dB will again be used. Connector losses are again allocated 6 dB (total both ends). The pulse spreading is assumed to be due to modal delay spread and is assumed to be 1 ns/km. Note that the spectral width (full width to half-maximum) of the LED at room temperature (and 1.3 μm wavelength) is about 100 nm. Our assumptions about pulse spreading imply that the dispersion of the fiber in the band of wavelengths being used is below 10 ps/km-nm. Based on these assumptions we can calculate the allowable distance of transmission (without repeaters) as a function of the data rate and based on loss and pulse spreading limitations. The results are shown in Figure 7-12. We see, for example, that at 25 Mb/s the spacing is limited by loss to 7—10.5 km. This is comparable to the spacings obtained with a laser and an APD at short wavelengths (see Figure 7-10). We show the curves as broken lines for data rates beyond 100 Mb/s because it is difficult to modulate typical long wavelength LEDs at these speeds. However, higher bandwidth LEDs have been reported. (140 Mb/s systems employing LEDs have been fielded in Europe.) The ability of the long wavelength LED/*p-i-n*/Multimode fiber system to provide long repeaterless spans, high data rates, high reliability, and high temperature tolerance have resulted in its popularity for many digital trunking applications.

7.2.4 Long Wavelength/Multimode/Laser/*p-i-n* Systems

As in the short wavelength case, we can obtain longer repeaterless spans and higher data rates of operation by using a laser transmitter instead of an LED.

The characteristics of the laser transmitter are similar to those given for short wavelength lasers (see Section 7.2.2 above) with the following exceptions.

The long wavelength laser threshold is more sensitive to temperature changes, which increases the difficulty of proper biasing somewhat.

The long wavelength laser is implemented with lattice matched layers. This increases its reliability at room temperature by about 1 or 2 orders of magnitude compared to GaAlAs lasers (for properly fabricated devices). However, the reliability of the long wavelength laser deteriorates more rapidly at elevated temperatures. Thus some ambient control is usually required.

If the laser is used at or near the wavelength of minimum dispersion, then mode partition noise is much less of a problem (compared to 0.8–0.9 μm operation). On the other hand, modal noise is still a problem when lasers are used with multimode fibers.

Long wavelength lasers can be modulated at higher data rates than short wavelength devices of similar dimensions, because the spontaneous carrier lifetime of the long wavelength material is about 2-3 times smaller

LASER OUTPUT: 0 dBm AVERAGE INTO A 50 μm CORE, 0.2 NA FIBER

RECEIVER SENSITIVITY: $-119 + 10 \log B$ (dBm)

MARGIN: 3 dB OR 10 dB

FIBER + SPLICE LOSS: 2 dB/km

CONNECTOR LOSSES: 6 dB TOTAL

PULSE SPREADING: 1.0 nsec/km (FWHM) ⇒ MAX BIT RATE $= 5 \times 10^8 / L$

Figure 7-13 Sample System Parameters — Long Wavelength Laser System with *p-i-n* Detector and Multimode Fiber.

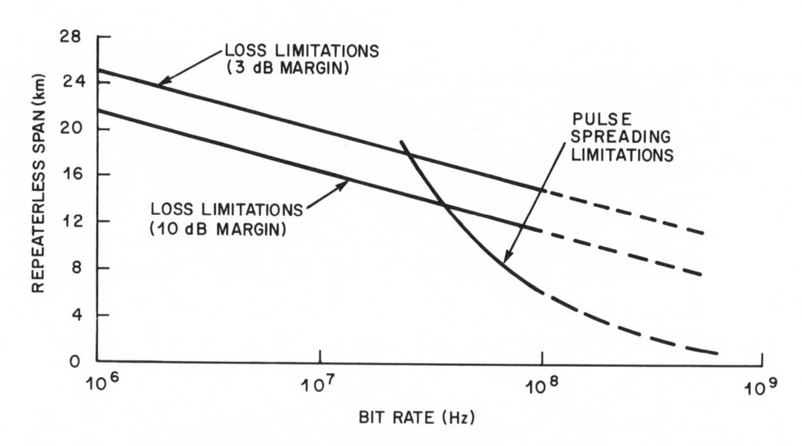

Figure 7-14 Repeaterless Span vs Data Rate — Long Wavelength Laser System with *p-i-n* Detector and Multimode Fiber.

than for GaAlAs. Other things being the same, the modulation frequency capability scales with the square root of this parameter.

Figure 7-13 shows a tabulation of typical system parameters for a long wavelength multimode fiber system with a *p-i-n* detector.

The parameters are the same as those given in the system example of Section 7.2.3 above, except that the transmitter output power has been increased to 0 dBm (average power launched into a multimode fiber). Figure 7-14 shows the allowable repeaterless span for a 10^{-9} bit error rate limited by loss and by pulse spreading.

We now begin to see some of the 15- to 20-km repeaterless spans achieved when systems of this type were deployed in the early 1980s.

7.2.5 Long Wavelength/Single Mode/Laser/*p-i-n* Systems

The use of a single mode fiber eliminates modal delay spread, significantly reduces (or eliminates) modal noise effects, and allows the system designer to take advantage of the lower losses of single mode fibers.

LASER OUTPUT: −6 dBm AVERAGE INTO A SINGLE MODE FIBER

RECEIVER SENSITIVITY: −119 + 10 LOG B dBm

MARGIN: 3 dB OR 10 dB

FIBER + SPLICE LOSS: 0.5 dB/km

CONNECTOR LOSSES: 6 dB TOTAL

MODE PARTITION NOISE LIMIT ⇒ MAX BIT RATE = $1.43 \times 10^{10}/L$

Figure 7-15 Sample System Parameters — Long Wavelength Laser System with *p-i-n* Detector and Single Mode Fiber.

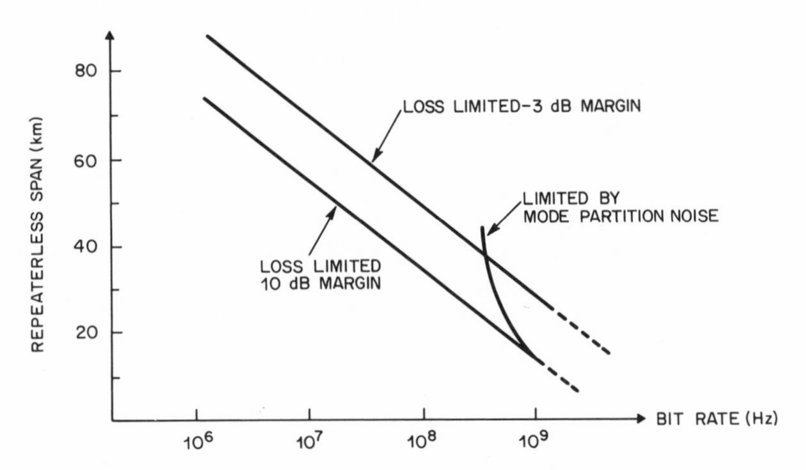

Figure 7-16 Repeaterless Span vs Data Rate — Long Wavelength Laser System with *p-i-n* Detector and Single Mode Fiber.

Figure 7-15 shows a tabulation of typical system parameters for a long wavelength single mode system using a laser source and a *p-i-n* detector.

The parameters are the same as those used in Section 7.2.4 above, except for the following. The cabled and spliced fiber loss has been reduced to 0.5 dB/km. The average power coupled into the single mode fiber by the laser has been reduced to −6 dBm from the 0 dBm level used in the multimode case. This accounts for the difficultly of coupling lasers to single mode fibers and compromises made to minimize reflections from the fiber into the laser.

Pulse spreading is now limited by material dispersion and manifests itself as mode partition noise rather than the classical intersymbol interference. In this example, we assume that the significant laser modes occupy a band of wavelengths 4 nm wide (about 9 modes with average power less than 20 dB below the dominant mode). This implies that the spread in delay is below 14 ps/km (assuming less than 3.5 ps/nm-km dispersion). We allow this spread in delay to accumulate to no more than 20% of the spacing between pulses. Thus, in this example, we have the following restriction on the repeaterless span, set by mode partition noise:

$$14 \times 10^{-12} \times L \times B \leqslant 0.2 \qquad (7\text{-}3)$$

where L is in km and B is in Hz. Figure 7-16 shows the allowable repeaterless span based on loss limits and mode partition noise limitations calculated from these parameters.

We see repeaterless spans of 22—36 km at a 400 Mb/s data rate. Mode partition noise begins to limit the repeaterless span at 350 Mb/s data rate. Note that the assumed value of margin is increasingly important as the loss per kilometer of the installed cable reaches these low levels. Also note that splicing and connector loss are similarly of increasing importance when the installed fiber loss becomes very low.

The use of single longitudinal mode lasers (when available) will eliminate mode partition noise. This will allow operation at higher data rates and at wavelengths of higher dispersion. However, direct modulation of laser diodes can cause the laser wavelength to sweep (chirp) over a wide enough range to cause delay distortion problems in very-long-distance high data rate systems operating at wavelengths where the dispersion is relatively high (e.g., 1.55 μm). Thus single frequency lasers may not be the complete answer to very-high-speed, very-long-distance operation.

In 1984 single mode systems operating at 400 Mb/s, at 1.3 μm wavelength, with 25- to 35-km repeaterless spans, were put into commercial service in the United States. Similar systems were installed in Europe and Japan on an experimental/or in-service basis previously. Some of these systems employed germanium APD detectors rather than InGaAsP *p-i-n* detectors, but the receiver performance was very similar in either case. The TAT8 undersea cable system planned for turn-up in 1988 will use 1.3 μm lasers and *p-i-n* detectors and operate at a data rate of several hundred Mb/s. The repeaterless span is anticipated to be between 35 and 65 km, depending upon the state of the art of fiber loss at the time the system is manufactured.

7.2.6 Long Wavelength/Single Mode/Laser/APD Systems

Recently long wavelength APDs, made from InGaAsP, with good multiplication statistics (but not as good as silicon) and very fast response speeds have been reported. Using such a device one can obtain about 5—10 dB of improved receiver sensitivity, which translates into 10—20 km of allowable distance, for an assumed installed fiber cable loss of 0.5 dB/km. (Provided one is not limited by mode partition noise.) Using such a device, researchers at AT&T Bell Laboratories were able to obtain a 203-km (125-mile) repeaterless span with a single frequency laser modulated at 420 Mb/s and emitting at 1.55-μm wavelength into a single mode fiber. The system operated with a bit error rate of 10^{-9} but did not have any connectors or margin. A repeaterless span of 130 km was obtained at 2 Gb/s data rate with similar components.

Long wavelength germanium APDs have been used by some manufacturers to obtain a few decibels of improved receiver sensitivity (vs *p-i-n* detectors) or to allow for more flexible tradeoffs in the design of the receiver preamplifier.

7.2.7 Long Wavelength/Single Mode/LED/*p-i-n* Systems

Recently, interest has emerged in the use of long wavelength LEDs, as an interim source, with single mode fibers. Although an LED can typically couple only about a microwatt of power into a single mode fiber, distances of more than 5 km have been achieved at data rates of 140 Mb/s. The underlying interest is in upgrading the system to a laser source for higher data rate operation at a future date, without changing the fiber.

7.3 Illustrations of Actual Equipment [11, 24]

It is helpful to present some photographs of actual equipment in order to lend reality to this discussion. Figure 7-17 shows a photograph of a research model (vintage 1974) of a 6.3-Mb/s one-way digital repeater employing a 0.85-μm LED and a silicon APD. The incoming fiber enters from the upper left and terminates on the detector. The detector is packaged in its own header next to a preamplifier, which uses a silicon FET for

Figure 7-17 Research Prototype 6.3-Mb/s Repeater (Photo). Circa 1974. Reprinted with Permission of Bell Communications Research and AT&T Bell Laboratories.

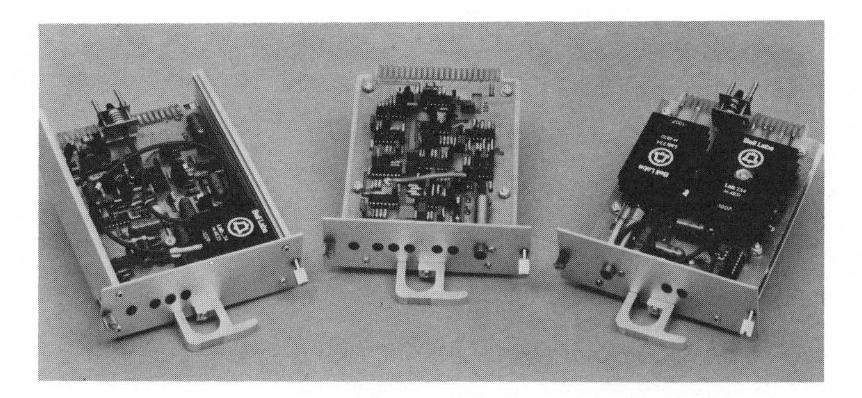

Figure 7-18 Experimental 44.7-Mb/s Repeater (Photo). Circa 1975-1977. Reprinted with Permission of Bell Communications Research and AT&T Bell Laboratories.

the first stage of gain. Below the preamplifier board are further stages of electronic gain. On the right side of the box, starting at the bottom, are band-limiting filter and clock recovery (phase locked loop) circuitry. The clocked regenerator and driver circuitry are at the top. The LED can be seen on the outside of the box at the top right.

Figure 7-18 shows a photograph of a 44.7-Mb/s repeater (vintage 1975—77) which was used in the Bell Laboratories Atlanta Experiment and later in the Bell System Chicago Project. The repeater is packaged on three boards and employs a GaAlAs (0.82-μm) laser and a silicon APD. The board on the right contains the laser, driver circuitry, and bias control circuitry to track the temperature dependence of the laser threshold (as described in Section 3.3.2 above). The laser is packaged in the module on the right side of the board along with a local detector to monitor its output and an attached fiber pigtail. The pigtail emerges from the module inside of a relatively large protective cable and terminates at the back of the board on a plug-in optical connector (as described in Section 2.4 above). When this board is inserted into a shelf, the electrical and optical connections are made automatically at the backplane. The board at the left contains the APD, the preamplifier, further stages of electronic gain, and a dc—dc converter for the APD bias. Also included is a peak-to-peak detector and a feedback control system of the kind described in Section 4.2.2 above. The APD and the preamplifier are integrated using chip devices on a thin film circuit board in the module at the lower right, along with an attached fiber pigtail. As in the transmitter, the pigtail emerges as a connectorized armored cable. The board in the middle contains only electronic circuitry for clock recovery and regeneration.

Figure 7-19 shows a photograph of a production model (circa 1979) of a 44.7-Mb/s repeater (now on one large plug-in board) which uses a GaAlAs laser and a silicon APD. The copper board at the front center contains a packaged laser module (out of view) and the laser bias control circuitry. A large heat sink for the laser is visible on the front panel. The APD and its preamplifier are contained in a small metal package barely visible at the lower left (behind the front panel). The input and output fiber connectors are visible on the backplane. A dc—dc converter is contained in the module at the upper left to generate high voltage for the APD bias. Amplifier and timing recovery circuitry occupies most of the real estate on the main board. A daughter board at the back of the main board contains fault location circuitry for maintenance.

Figure 7-19 Production Model of a 44.7-Mb/s Repeater (Photo). Reprinted with Permission of AT&T Bell Laboratories.

Figure 7-20 shows an unpackaged receiver module made from chip components integrated on a thin film hybrid substrate. The object at the upper right is a chip capacitor to decouple the bias for the detector. A silicon APD mounted on a small ceramic block is located to the left of this capacitor. A circular metal contact surrounds the light sensitive area. Chip transistors (4) are visible on the board. Figure 7-21 shows a

Figure 7-20 Experimental Receiver Module (Photo). Courtesy of TRW Inc.

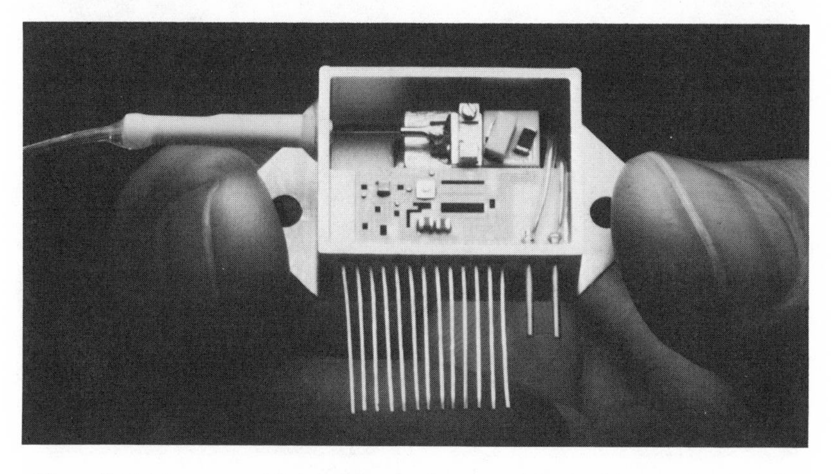

Figure 7-21 Transmitter Module (Photo). Courtesy of Rockwell International, Collins Transmission Systems Division, Dallas, Texas.

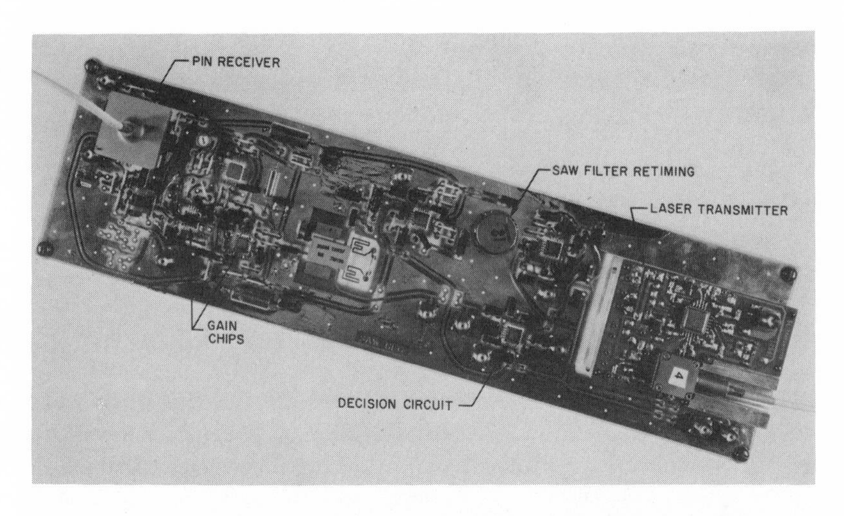

Figure 7-22 Experimental Prototype of an Undersea Cable Repeater (Photo). Reprinted by Permission of AT&T Bell Laboratories.

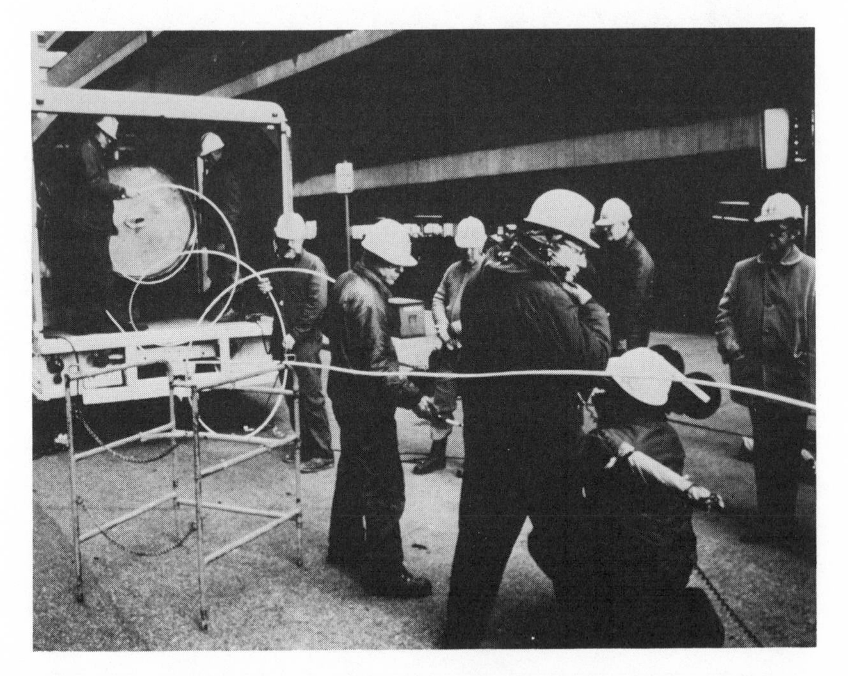

Figure 7-23 Cable Installation. Chicago 1976. Reprinted by Permission of Bell Communications Research and AT&T Bell Laboratories.

transmitter module including a laser (at upper center), a local detector (on an angled block to the right of the laser), laser bias feedback control circuitry, driver circuitry, and an attached fiber pigtail (fiber visible at upper left).

Figure 7-22 shows a prototype of the repeater which will be part of the proposed TAT8 undersea cable system [24, 25]. The detector and its preamplifier are packaged in the metal module at the left. The laser and its local monitor detector are packaged in the metal module at the right. High speed monolithic integrated circuits perform the amplification, regeneration, laser control, laser driver, and maintenance functions. The clock recovery circuit employs a surface acoustic wave (SAW) filter.

Figure 2-23 shows a diagram of a cable containing 24 multimode fibers (as described in Section 2.3 above). Figure 7-23 shows a cable of this type being installed in an underground duct during the installation of the Bell System Chicago Project in early 1977. The cable is being pulled through a hose (inner duct) which was itself pulled through the duct at an earlier date. To minimize pulling tension in this early experiment, the

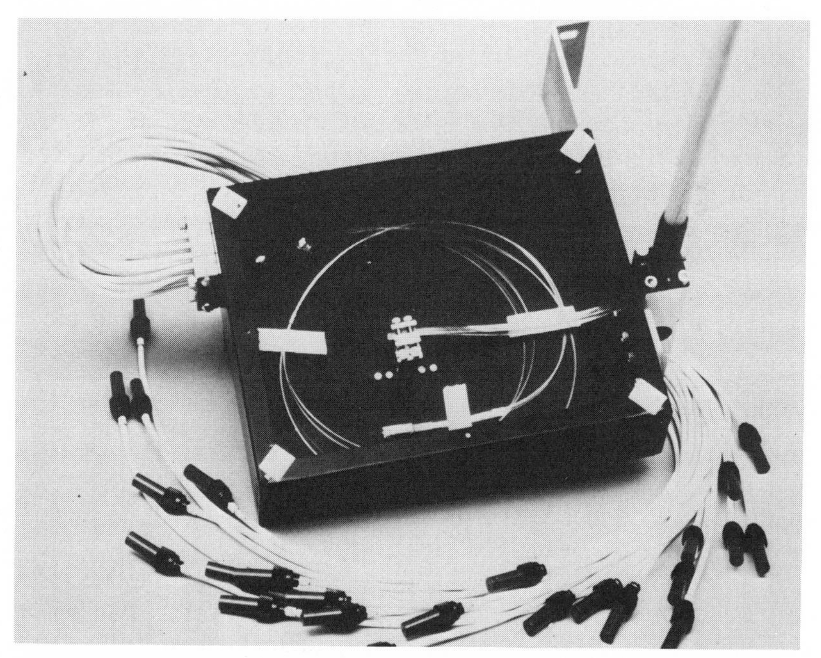

Figure 7-24 Fanout Box (Photo). Chicago 1976. Reprinted by Permission of Bell Communications Research and AT&T Bell Laboratories.

Figure 7-25 Equipment Room (Photo). Chicago 1976. Reprinted by Permission of Bell Communications Research and AT&T Bell Laboratories.

cable was pulled in both directions from an intermediate manhole location. For this reason, in one direction of the pull, the cable must be taken from the side of the reel as shown. It untwists during the pulling operation.

Figure 7-24 shows a fanout box which allows the cable, containing two ribbons of 12 fibers each, to interface to 24 individually armored and connectorized pigtails (shown protected with dust caps). The ribbons emerging from the cable are terminated in a 2 × 12 array connector (as described in Section 2.4 above). The individual fibers that enter the box are combined into two ribbons and terminated with a matching connector. The connectors are mated in the fixture at the center of the box.

Figure 7-25 shows a room filled with racks of channel banks, multiplexing equipment, back-up power supplies (storage batteries), protection switching equipment, etc. which was part of the Chicago Project. The optical transmitters and receivers and an optical patch panel are contained in the rack at the far right.

Figure 7-26 shows the shelf containing two terminal repeaters. One is the main repeater and one is a standby repeater which is automatically switched into service (if a failure is detected) by protection switching equipment. Each terminal repeater consists of three plug-in boards as shown in Figure 7-18.

Figure 7-26 Repeater Shelf (Photo). Chicago 1976. Reprinted by Permission of Bell Communications Research and AT&T Bell Laboratories.

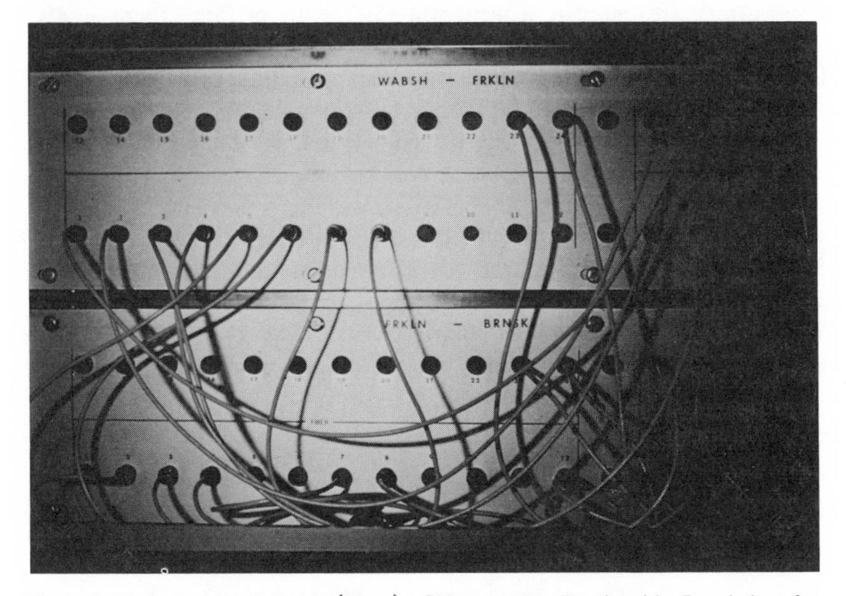

Figure 7-27 Optical Patch Panels (Photo). Chicago 1976. Reprinted by Permission of Bell Communications Research and AT&T Bell Laboratories.

Figure 7-27 shows two optical patch panels which terminate the pigtails from two 24 fiber fanouts, and the backplane jumper cables running from the backplanes of the repeater shelves.

A connection between any two sockets is established with an optical jumper cable having plug-in connectors on both ends.

It is interesting that in this system, operating at 0.82 μm wavelength, one could see the "red" light eminating from sockets in the patch panels which were illuminated by light from distant lasers.

7.4 Wavelength Multiplexing Applications

The use of two or more wavelengths, independently, in the same fiber, can be accomplished with wavelength selective components. Examples of such components were described in Section 5.1.2 above. Figure 7-28 illustrates the wavelength multiplexing scheme called diplexing, where two wavelengths are transmitted in the same direction over a common fiber. We assume that wavelength 1 is a shorter wavelength, and that wavelength 2 is a longer wavelength. A wavelength multiplexer combines the light from the two transmitters and couples the composite light signal into a single outgoing fiber. In theory, if the two wavelengths have no overlapping spectral components, this can be done in a lossless manner. That is, the coupling efficiency obtained for each of the transmitters to the fiber, when coupled simultaneously, can be as good as when the transmitters are coupled to separate fibers. In practice there is always some extra loss associated with the use of a wavelength multiplexer (beyond the coupling loss each wavelength would experience if coupled separately, without the multiplexer), but this penalty in coupling can be kept below 1−2 dB with proper coupler design (even for single mode fibers). It is also possible to use a non-wavelength-sensitive coupler of the type discussed in Section 5.1 above. However, in that case, there will be an additional 3 dB of coupling loss at each wavelength (if equal coupling is desired), unless the transmitters are coupled into different mode groups of a common multimode fiber (including the possibility of coupling two transmitters into orthogonal polarizations). The two wavelengths propagate through the fiber, possibly experiencing significantly different losses.

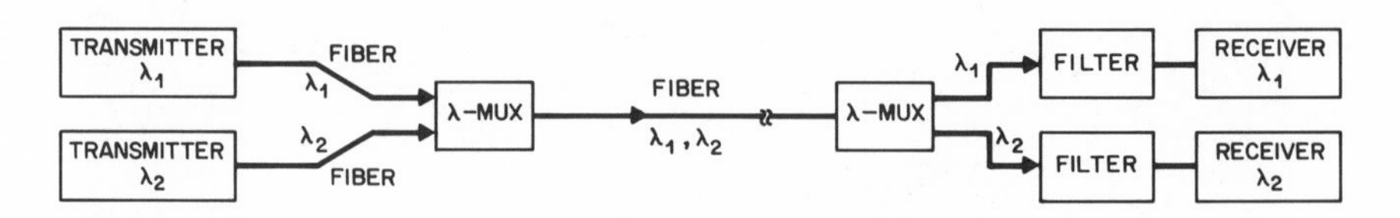

Figure 7-28 Diplexing with WDM Technology.

At the opposite end of the fiber, the wavelengths are partially separated by a wavelength selective demultiplexer. To obtain adequate rejection, at the inputs to the receivers, of the unwanted wavelengths, additional long-pass and short-pass wavelength filters may be required. The wavelength demultiplexer introduces some excess loss, just as in the case of the wavelength multiplexer; but this can be kept below 1–2 dB if the wavelength selectivity required is not too great. Similarly, the long pass and short pass filters introduce some loss, depending upon the selectivity required. The rejection of the unwanted wavelength required of the combination of the wavelength demultiplexer and the wavelength selective filters depends upon a number of factors. The two optical signals arriving at the end of the fiber may not be of equal level, for three reasons. The two transmitters may emit unequal levels of optical power. The coupling efficiency of the transmitter output signals to the fiber may not be the same for both wavelengths. The fiber loss may be different for the two wavelengths. Thus one of the wavelengths may arrive at the demultiplexer at a significantly lower optical level than the other. This will increase the required rejection of the higher level optical wavelength in the path leading to the receiver intended for the lower level optical wavelength. If digital modulation techniques are used then a ratio of 10:1 between the power levels of the desired and the interfering optical signals at the receiver input may be adequate for error-free detection of the desired signal. If analog modulation techniques are being used, then much higher ratios of desired-to-interfering signals may be required to achieve the desired signal-to-noise (interference) ratio at the receiver output.

Additional rejection of the undesired wavelength signal may be provided by the detector if the wavelengths are separated sufficiently to experience different responsivities. For example a silicon detector will not respond significantly to 1.3-μm light, but will have a 70%–80% quantum efficiency in the 0.8–0.9 μm band of wavelengths.

Example

In a diplexing system, laser sources are used at 1.275- and 1.325- μm wavelengths. The laser transmitters employ closed loop control to stabilize their output power levels to within ±0.5 dB of nominal. Each laser has a nominal output power of −5 dBm into an attached single mode fiber pigtail, and each is individually modulated with a 400 Mb/s digital signal.

The coupling loss from either output pigtail into a common single mode fiber (via a wavelength multiplexer) is −2.5 dB ± 0.5 dB (which includes a 0.5-dB nominal coupling loss of either pigtail to the fiber, if used without wavelength multiplexing). The fiber loss is 15 dB at the longer wavelength (nominally) and up to 3 dB higher at the shorter wavelength.

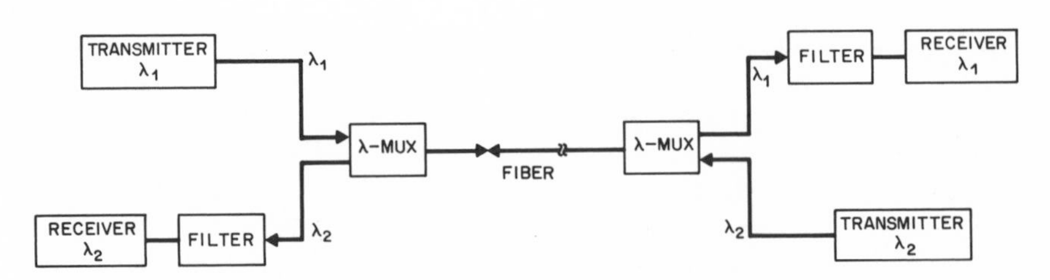

Figure 7-29 Duplexing with WDM Technology.

In the worst case combination of these parameters, the longer wavelength signal would arrive at level which is 5 dB stronger than the shorter wavelength signal.

The receiver requires a 10:1 rejection of the unwanted wavelength to avoid excessive performance degradation due to interference.

This implies that the wavelength demultiplexer and the short pass filter (if used) must have a combined contrast between the transmission of the short and long wavelengths of 15 dB or more.

Furthermore, if we allow 2 dB of total insertion loss for the demultiplexer and short pass filter (if used), then the signal level arriving at the shorter wavelength receiver could be as small as −28.5 dBm. The shorter wavelength receiver would have to be sensitive enough to accommodate this signal level, in the presence of an interfering signal level of up to −38.5 dBm.

Figure 7-29 illustrates the wavelength multiplexing scheme called duplexing, where two wavelengths are transmitted in opposite directions over a common fiber. The considerations here are similar to the diplexing case, with the following additions. The interference from the undesired wavelength arises from reflections from the wavelength multiplexing components, splices, and Rayleigh scattered light (see Section 2.1.3 above), which can be recaptured in the backward direction. These reflected signals can be relatively large, compared to the desired optical signal, if the desired signal has experienced substantial attenuation in propagating through the fiber, and if the reflections are near the transmitting end for the undesired signal. For this reason the required contrast between the transmission of the desired and the undesired optical signals by the multiplexers and optical filters may be higher in duplex systems than in diplex systems. For example a nearby reflection from a fiber splice which is not properly index matched can direct 8% (−11 dB) of the undesired wavelength signal back toward the wavelength multiplexer. Meanwhile, if the fiber loss is 30 dB at the wavelength of the desired signal, then it will arrive at the receiver from the far end at only 0.1% of its initial level

(−30 dB). To obtain a 10:1 contrast of the desired and undesired signals at the receiver will require a combined contrast in the transmission of the wavelength multiplexer and optical filter (if used) of at least 29 dB (30 dB + 10 dB − 11 dB).

Combinations of duplexing and diplexing with more than two wavelengths are also possible.

The usual motivation for wavelength multiplexing is to share the fiber cost amongst more information-bearing signals. However, there are some applications where one desires two-way communication; but where the use of more than one fiber is inconvenient. For example, there are certain compact cable designs which are difficult to implement with more than one fiber. As another example, most connectors are easier to implement with a single fiber; and sometimes multiple connectors are not convenient. A rotary joint is an example of an interconnection which would be much more complex (perhaps impossible for practical purposes) with two fibers than with one. (A few months after the preceding sentence was written, this author read a paper describing a practical two-fiber rotary joint! The lesson learned is that one should be cautious in speculating on what is impossible [33].)

The key to the successful implementation of a wavelength multiplexed system is obtaining and maintaining adequate isolation of the independent signals being transmitted and received.

A number of mechanisms can give rise to interference between the signals at nominally different wavelengths. The wavelength multiplexing optical components and wavelength selective filters are not perfect. They have a limited ability to discriminate between even widely spaced wavelengths (due to scattering mechanisms and other imperfections); and the ability to discriminate between closely spaced wavelengths is hampered by the sensitivities of filters and gratings to fabrication tolerances and temperature effects.

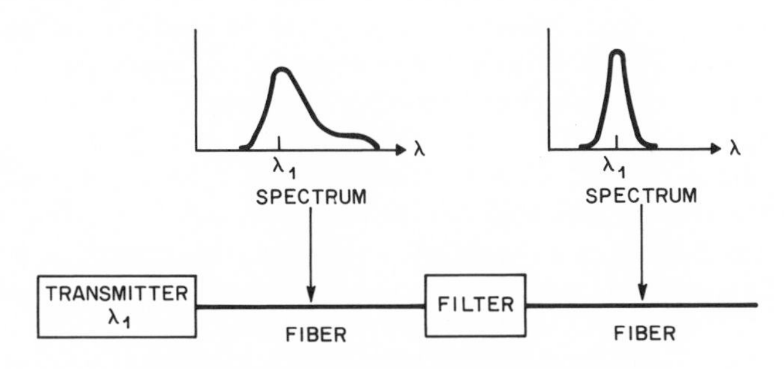

Figure 7-30 Bandlimiting a Broadband Source.

In addition to the problems of unwanted signal leakage through the nominal stop bands of wavelength multiplexers and filters, one has to take into account emissions of optical sources which may fall within the wavelength bands allocated to other sources. For example, a GaAlAs LED may have a nominal bandwidth of 50 nm (full width to half-maximum of its emission spectrum at room temperature). However, it may emit intolerable levels of power over a much wider range of wavelengths, depending upon the isolation required in the wavelength multiplexed system. To help reduce out-of-band emissions from a source one may have to include bandpass filters at the source as well as at the receiver as shown in Figure 7-30.

Full duplex single mode systems have been demonstrated, in the laboratory, using 1.3- and 1.5-μm wavelength laser sources and interference filters for wavelength selectivity, operating with digital modulation at 144 Mb/s over 58 km of fiber [34]. AT&T Technologies is considering the implementation of a three-wavelength multimode triplex system using two lasers in the 0.8—0.9 μm band of wavelengths and one LED at 1.3 μm wavelength. This could be used to enhance some of AT&T Communications' multimode fiber installations by permitting 3—90 Mb/s digital signals to be transmitted over the same fiber.

7.5 Interfacing with Synchronous Digital Signals

In the telecommunications trunking application, the terminals interfacing the optical transmission facility produce digital signals at a fixed standard rate, and, if necessary, can provide a clock signal at this same rate. A typical optical fiber transmission facility requires a signal which has a reasonably good balance in the number of ones and zeros in a time window of several hundred bit intervals, and which makes a reasonable number of transitions from the one state to the zero state (and vice versa) in a similar interval of time. The balance is required to allow for ac coupled electronics and automatic gain control circuitry in the optical receiver (and possibly in the transmitter as well). The transitions are required to provide a capability for the receiver to recover a clock signal from the incoming data stream. If the terminal equipment (multiplexer or other digital signal source) does not produce a signal with guaranteed balance and transitions, then this can be accomplished by the use of scrambling or coding between the terminals and the transmission facility. The coder and decoder can be built into the optical transmitter and receiver, respectively.

The scrambler is not intended to provide security from eavesdropping, but rather a mechanism for making the bit stream more random (providing balance and transitions). Figure 7-31 shows a diagram of a simple scram-

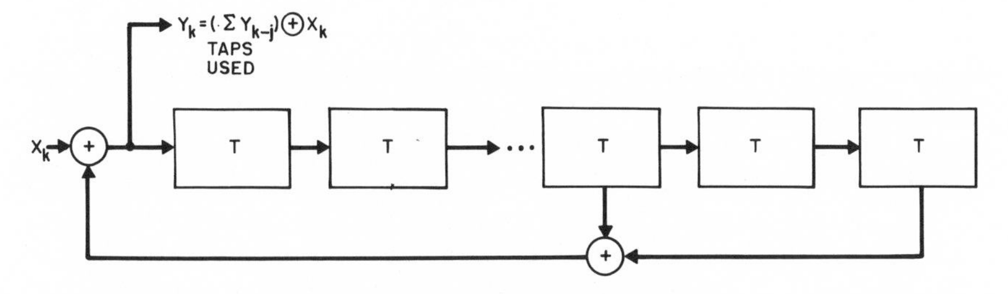

Figure 7-31 Block Diagram of a Scrambler.

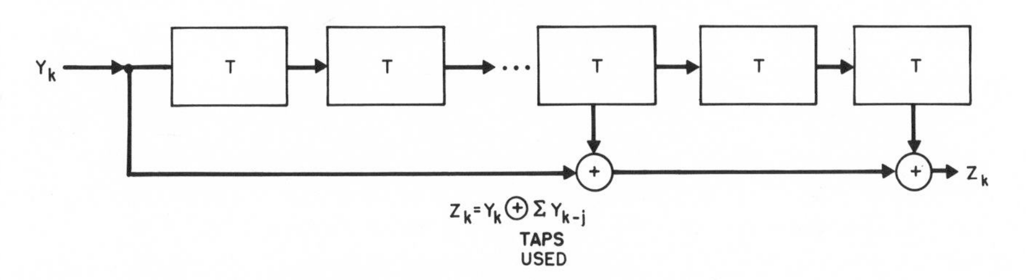

Figure 7-32 Block Diagram of a Descrambler.

bler, along with a formula relating the output of the scrambler to the input. The symbol consisting of a plus sign inside of a circle stands for modulo 2 addition. Figure 7-32 shows a diagram of the complementary descrambler, along with its output–input relationship. By simple modulo 2 algebra we can see that the output of the descrambler is identical to the input to the scrambler:

$$Z_k = Y_k \oplus \sum_{\text{Taps used}} Y_{k-j} = X_k \oplus \sum_{\text{Taps used}} Y_{k-j} \oplus \sum_{\text{Taps used}} Y_{k-j} = X_k \tag{7-4}$$

Scrambling does not increase the data rate of a signal, but it also does not provide as good a control of dc balance and transition density as coding.

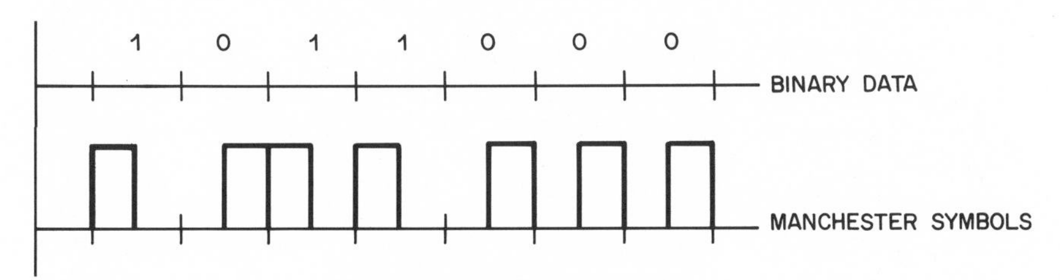

Figure 7-33 Manchester Coding.

Figure 7-33 illustrates a coding technique called Manchester coding, where each on/off binary coded symbol is converted to a pulse in either the left or the right half of a corresponding bit interval. Since each Manchester bit interval contains a pulse, the signal is balanced. Manchester coding has the effect of doubling the bandwidth required of the transmission facility. More sophisticated coders code n binary symbols into a larger number of binary symbols, m, such that the data rate of the coded signal is m/n times the original signal rate. The extra bits introduced in the coding process allow for the selection of groups of m bits (code words) which are either dc balanced or which have unbalance to compensate for previous unbalanced groups. Manchester coding is a member of the "n on m" family of coding techniques where $n = 1$ and $m = 2$.

Problems

1. A transmission link has the following parameters:

   a. Source: 1.3 μm laser; −6 dBm into a single mode fiber (average power); 10:1 on−off ratio (energy in "zero" ÷ energy in "one" = 0.1)

   b. Receiver: 1.3 μm p-i-n detector; sensitivity = 12,000 photons per received (on) pulse

   c. Installed single mode cable loss (with splices): 0.5 dB/km

   d. Margin required: 6 dB

   e. Connector loss (total of both ends): 4 dB

   f. Fiber dispersion: 3.5 ps/km-nm

   g. Full width of laser optical spectrum: 4 nm

   h. Maximum allowable delay difference between slowest and fastest wavelength: 0.2 × pulse spacing (pulse spacing = 1/bit rate)

   Calculate the allowable transmission distance (km) without a repeater for both loss-limited and mode-partition-noise (delay) limited conditions; and plot (km vs B) for $10 \leqslant B \leqslant 1000$ Mb/s.

2. A transmission link has the following parameters

   a. Source: 1.3 μm LED; −18 dBm into a 50 μm core, 0.2 N.A. fiber (average); 10:1 on−off ratio

b. Receiver: 1.3 μm p-i-n detector; sensitivity 12,000 photons per received (on) pulse

c. Installed cable loss: 1.5 dB/km with splices

d. Margin required: 6 dB

e. Connector losses: 4 dB (total)

f. Pulse spreading due to mode delay differences: 2 ns/km

g. Pulse spreading due to material dispersion: 0.5 ns/km

h. Maximum allowable *total* pulse-spreading: 0.5 × pulse spacing (1/bit rate)

Reminder: (total pulse-spreading)^2 = (mode delay spreading)^2 + (dispersion spreading)^2

Calculate and plot the allowable repeaterless span (km) vs B, for $1 \leqslant B \leqslant 200$ Mb/s, for loss-limited and for pulse-spreading-limited conditions.

3. Repeat Problem 2, except:

   a. Fiber is single mode, cable loss is 0.5 dB/km with splices, no pulse spreading due to mode delay differences (material dispersion still present)

   b. Launched power into single mode fiber from LED is −33 dBm (average).

8

Data Links

It is often necessary to provide a connection between two computing devices which can be anywhere from a few inches to a few kilometers apart. For the purposes of this discussion, we define computing devices as electronic terminals which exchange signals in digital form. Figure 8-1 shows an example, where a piece of digital terminal equipment (e.g., a computer terminal) communicates with a piece of digital communications equipment (e.g., a modem) using a fiber link. Figure 8-2 shows a different example, where a piece of digital terminal equipment (DTE) communicates with a piece of digital communications equipment (DCE) using a pair of wires, but the digital communications equipment communicates to another piece of digital communications equipment using a fiber link.

Why would we use fiber to implement a short link? There are several possible reasons. As the clock rates of microprocessors and other

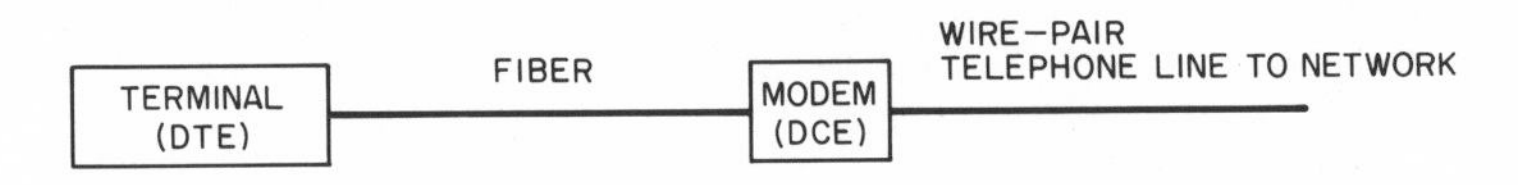

Figure 8-1 DTE-DCE Fiber Data Link.

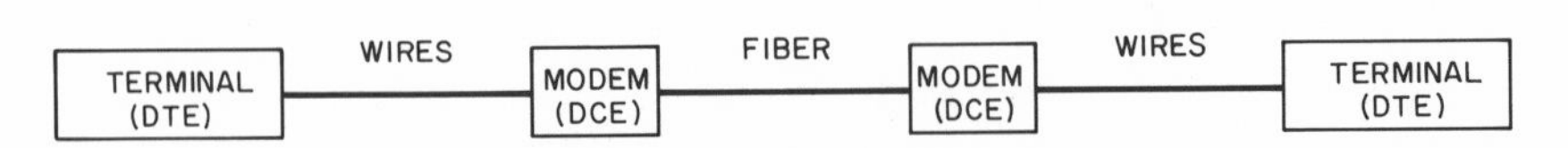

Figure 8-2 DCE-DCE Fiber Data Link.

computer components increase, it is becoming increasingly difficult to interconnect these components with simple twisted pairs (even over short distances) because of practical problems like crosstalk between pairs, impedance mismatches at connectors, frequency dependent attenuation, etc. Twisted pairs (and even coaxial cable connections) carrying high frequency digital waveforms tend to radiate unacceptable amounts of interference. This has become increasingly problematic with the proliferation of consumer products containing digital electronics. Similarly, in addition to causing unwanted interference, this radiation provides increased opportunities for eavesdropping.

All of these problems can be solved with properly designed fiber optic interconnections.

On the other hand, to implement even short point-to-point links, one still requires an optical transmitter and an optical receiver to interface with the conventional electronics at each end. For short links, of moderate clock rates, these subsystems must be inexpensive, reliable, low in power consumption, and must operate from conveniently available power supply voltages.

Furthermore, they must be convenient to use (e.g., simple rugged optical connectors).

We can point out some important contrasts between the data link application and the point-to-point telecommunications trunking application described in Chapter 7 above.

A data link tends to be much shorter in length than a telecommunications trunk. The transmitter and receiver of a data link must typically be substantially less costly and substantially smaller than for a telecommunications trunk. The required reliability (average failure rate) is often higher for a data link (but not always). The range of ambient temperatures which must be tolerated may be larger for the data link transmitter and receiver. The signal arriving at the data link input is often a limited duration packet of digital bits with no specific timing relationship to previous packets (asynchronous).

These differences have a number of implications, for example, the shorter length of a data link allows for the use of higher loss fibers and lower output power transmitters, in trade for lower cost, higher reliability, ease of making optical connections, etc. The required higher reliability and wider range of ambient temperatures make LED sources preferable to laser diode sources. The need to interface asynchronous signals leads to substantial differences in the way signals are coded and/or the way receivers are designed, compared to the telecommunications application (where the information is a continuous stream of data at a constant clock rate).

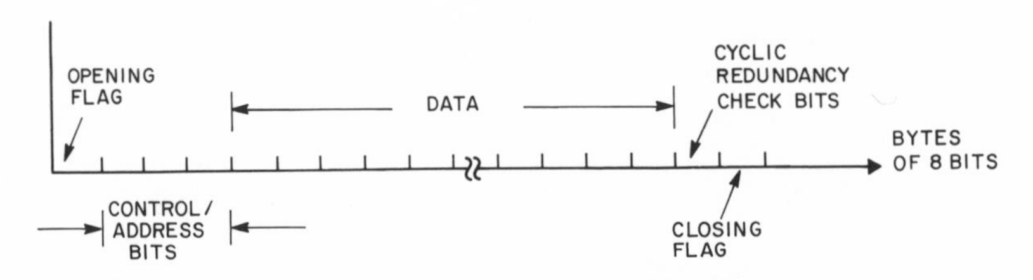

Figure 8-3 Hypothetical Asynchronous Data Packet.

8.1 Interfacing Asynchronous Signals

In the telecommunications trunking application, the signal arriving at the input to a fiber transmission facility is a continuous sequence of binary ones and zeros, at a nearly constant (and well specified) clock frequency. As described in Section 7.5, about the only practical concerns one typically has with such a signal is whether or not there is a good balance of ones and zeros (i.e., the signal can be ac coupled without causing base line wander) and whether there are enough transitions between the two states to provide good clock recovery from the received signal. In the trunking application, balance and adequate density of transitions can be guaranteed by well-known coding techniques, which can be implemented in the terminals generating the signals, or as part of the fiber optic transmitter and receiver.

Signals originating in computing terminal equipment are often in the form of asynchronous packets of digital bits. The first transition in the packet may occur at an arbitrary time (no clock reference), and the number of bits in the packet may be either fixed in advance or variable from packet to packet. Typically the spacing of allowable transition times within a packet (the internal clock rate of the packet) is the reciprocal of one of a number of standard rates (e.g., 9.6 kbaud) and is accurate to a tight tolerance.

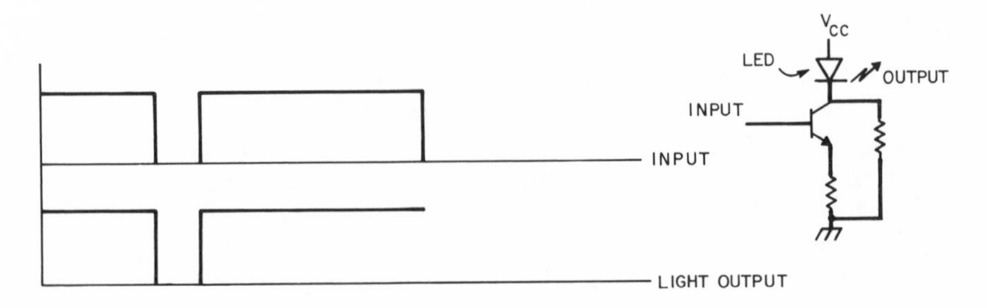

Figure 8-4 Direct Modulation Asynchronous Interface Transmitter.

Figure 8-3 shows a sample of a sequence of binary electrical symbol timeslots comprising a hypothetical packet of digital information.

Figure 8-4 shows one straightforward way to convert binary electrical symbols to optical form, called direct modulation. The optical transmitter simply turns on and off in response to the electrical input signal. The advantage to this approach is that any input signal can be accommodated, provided its rise and fall times are within the capabilities of the transmitter, the fiber, and the optical receiver at the other end of the link. A disadvantage of this approach is that the transmitter, and the receiver at the opposite end of the fiber link, must both be dc coupled (or equipped with dc restoration), since there is, in general, no guarantee of a long term balance between the off state and the on state. Another disadvantage is that the transmitter can be in the off state for long periods of time. There is no guaranteed optical signal at the receiver which can be used for link performance monitoring or for setting a slow-acting automatic gain control. One could attempt to provide a signal to the receiver, for the purpose of link performance monitoring (verifying that the transmitter and the fiber are functioning) and for adjusting a slow-acting automatic gain control, by setting the transmitter output level to a nonzero value in the "off" state. A difficulty with this approach is that the receiver must have a large enough optical signal in the off state to distinguish that signal from its internal electronic dc offsets.

The receiver can be designed to simply produce an electrical "high" output level if the input optical signal is above a given threshold, and an electrical "low" output level if the input optical signal is below a given threshold. What we then end up with is a simple optical link, suitable for short distances and moderate data rates. Such a receiver is shown in Figure 8-5. The short distance limitation is imposed by the fact that the signal level at the receiver must be large enough to be unambiguously above or below the threshold level in the presence of variations in the optical loss between the transmitter and the receiver, dc offsets in the transmitter and

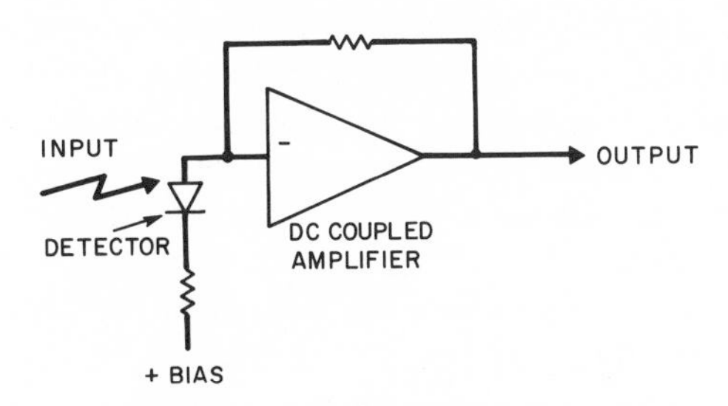

Figure 8-5 Direct Modulation Receiver.

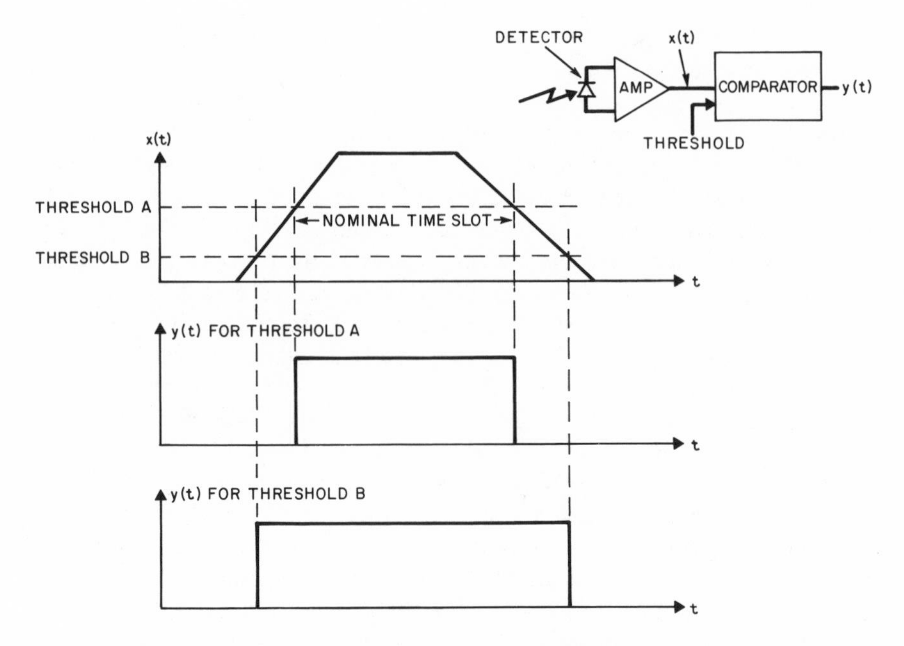

Figure 8-6 Data Rate Limitations Caused by Threshold-Crossing-Time Ambiguities.

receiver electronics, and noise within the full band of frequencies passed by the receiver. The moderate data rate limitation is imposed by ambiguities in the time at which the optical signal arriving at the receiver will cross the threshold level, caused by rise and fall time limitations of the electronic components, ambiguities in the arriving optical signal level, and noise. Figure 8-6 illustrates this problem.

One can implement a more sophisticated receiver as shown in Figure 8-7. Here the first few transitions at the beginning of an arriving packet are detected, and cause an initialization process to begin. The receiver attempts to estimate the level of the incoming packet from these first few arriving bits and fixes the setting of a variable gain amplifier to accommodate the level of the incoming optical signal for the remainder of the packet. The receiver uses the first few transitions in the incoming packet to set the phase of a clock (in the receiver) which is used to sample the incoming waveform for the remainder of the packet. If the clock frequency of the incoming packet has a tolerance of 0.0001% and the internal clock of the receiver has a similar tolerance, then once the phases are aligned, they will stay aligned for many tens of thousands of bits.

In order to implement the above fast-acting automatic gain control strategy, something must be known about the structures of the incoming

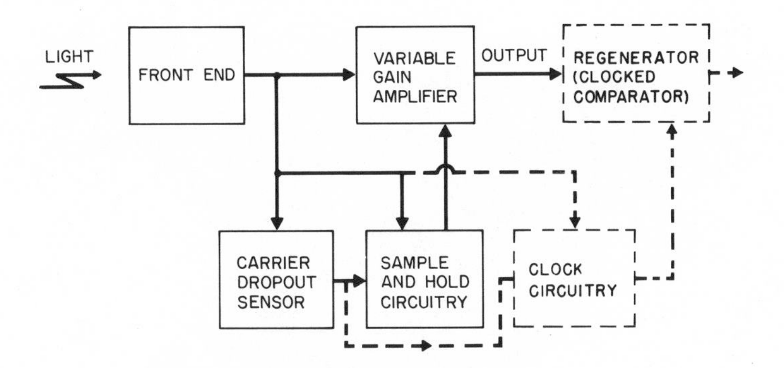

Figure 8-7 Fast Adaptation Receiver.

packets, and the packets must include a training sequence to accommodate the initialization procedure.

In order to implement the fast acting clock recovery strategy, the internal packet clock rate must be known and controlled to a high tolerance.

At the end of the packet, the receiver (carrier drop-out sensor) detects a long sequence of zeros, and prepares to initialize the next packet.

The advantage of this scheme is that the receiver can adjust to the level of the incoming signal, and can therefore accommodate smaller signals (is more sensitive). Furthermore, the receiver regenerates timing (via its internal clock), and therefore the link can operate at a higher rate without introducing unacceptable jitter in the zero crossing times at the receiver output relative to the transmitter input.

Figure 8-8 shows a strategy for interfacing an asynchronous digital signal to a fiber optic link, which is called oversampling. This approach is frequently used when a number of relatively low rate asynchronous signals are multiplexed together to interface a common digital fiber link (operating at its own fixed rate). Each asynchronous signal is sampled at a fixed rate

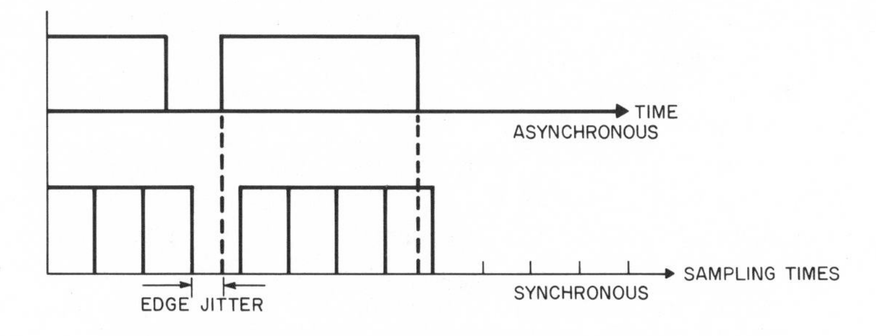

Figure 8-8 Oversampling as an Asynchronous Interface Strategy.

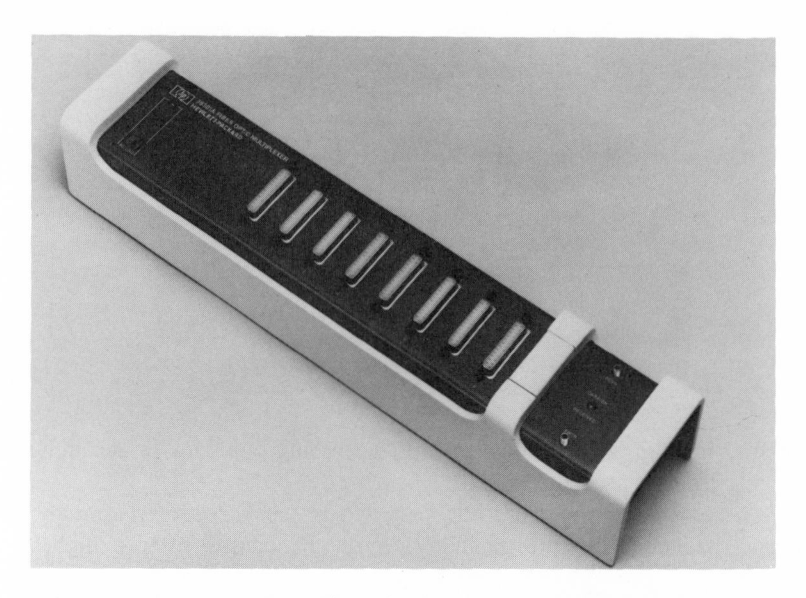

Figure 8-9 Fiber Optic Modem (Photo). Courtesy of Hewlett Packard Company.

(synchronized to the clock in the fiber transmission link), which is at least 10 times (typically) as fast as the baud of the asynchronous signal. If the asynchronous signal is in the high state when sampled, then a digital one is produced in the associated fixed rate waveform. If the asynchronous signal is low when sampled, then a digital zero is produced in the fixed rate waveform. There is some error introduced in this sampling process in reproducing the transition times of the asynchronous waveform. The allowable error determines the required ratio of the sampling rate to the asynchronous baud. Once the asynchronous waveform has been converted to a synchronous waveform (at a fixed clock rate and phase relationship to the clock in the fiber transmission link) it can be coded, multiplexed with other synchronous digital waveforms, and transmitted over a fixed-rate synchronous fiber link. Such a fiber link can incorporate ac coupled components and conventional slow acting automatic gain control and clock recovery strategies.

For example, if one wishes to interface eight asynchronous signals, each with a maximum baud of 9.6 kHz to a common fiber link, then one could sample each signal at 100 kHz, and end up with a modest composite (multiplexed) rate of 800 kHz. Figure 8-9 shows a photograph of a fiber optic modem which interfaces up to eight standard RS232 signals to a fiber link, using oversampling and synchronous multiplexing of the sampled waveforms.

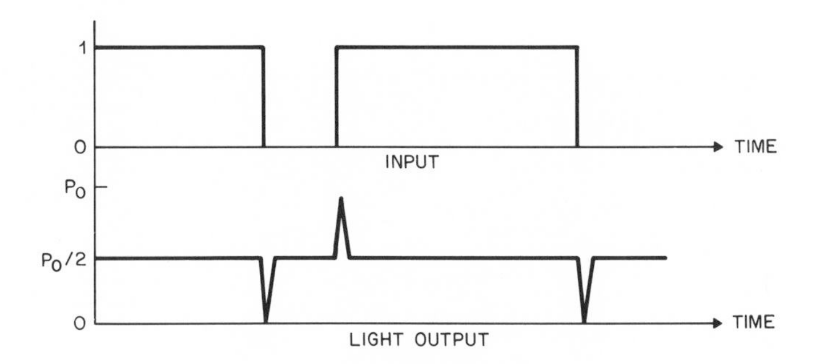

Figure 8-10 Transitional Coding Strategy.

Figure 8-10 shows a method of interfacing asynchronous waveforms called transitional coding. An optical transmitter which can produce three levels of output (off, half-on, and on) is required to implement this method. The optical transmitter produces a half-on output level, unless a positive-going or a negative-going transition occurs in the asynchronous waveform. A positive-going transition in the asynchronous waveform causes the optical transmitter to emit a short duration pulse of full output. A negative-going transition in the asynchronous waveform causes the optical transmitter to produce a short duration pulse of zero output. In addition, if the asynchro-

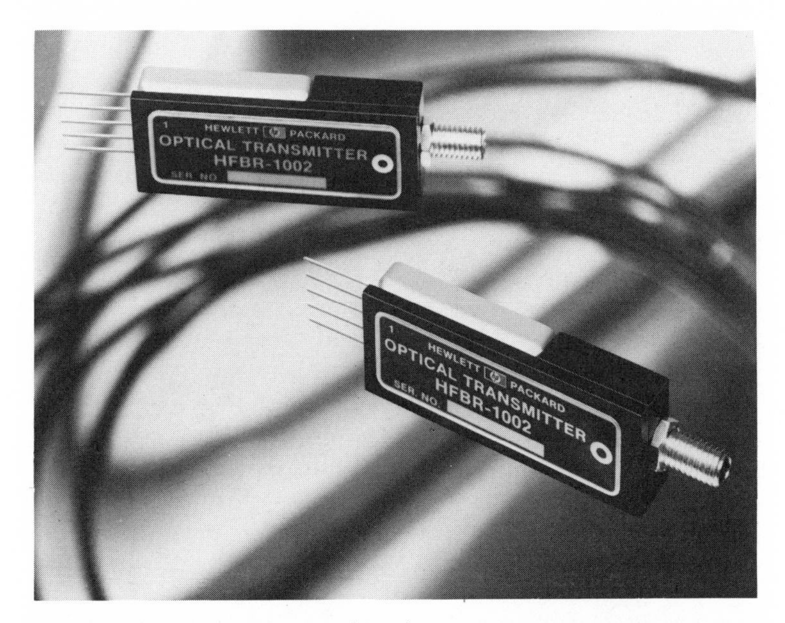

Figure 8-11 Data Link Transmitter (Photo). Courtesy of Hewlett Packard Company.

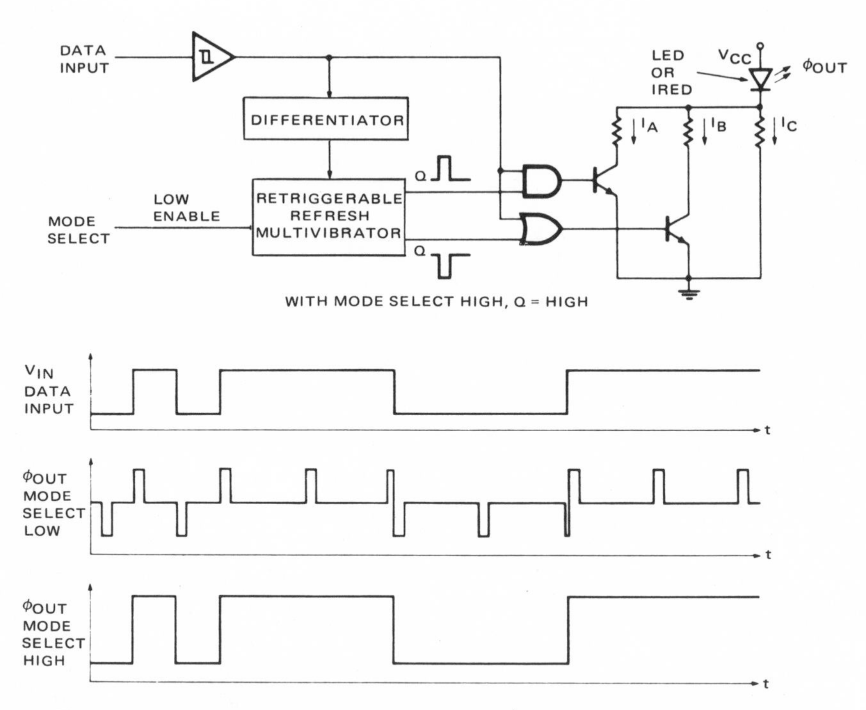

Figure 8-12 Schematic of Data Link Transmitter. Courtesy of Hewlett Packard Company.

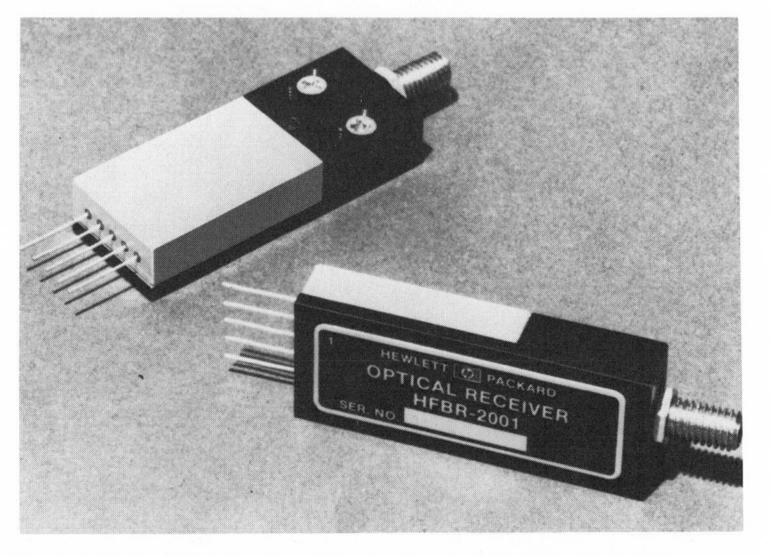

Figure 8-13 Data Link Receiver (Photo). Courtesy of Hewlett Packard Company.

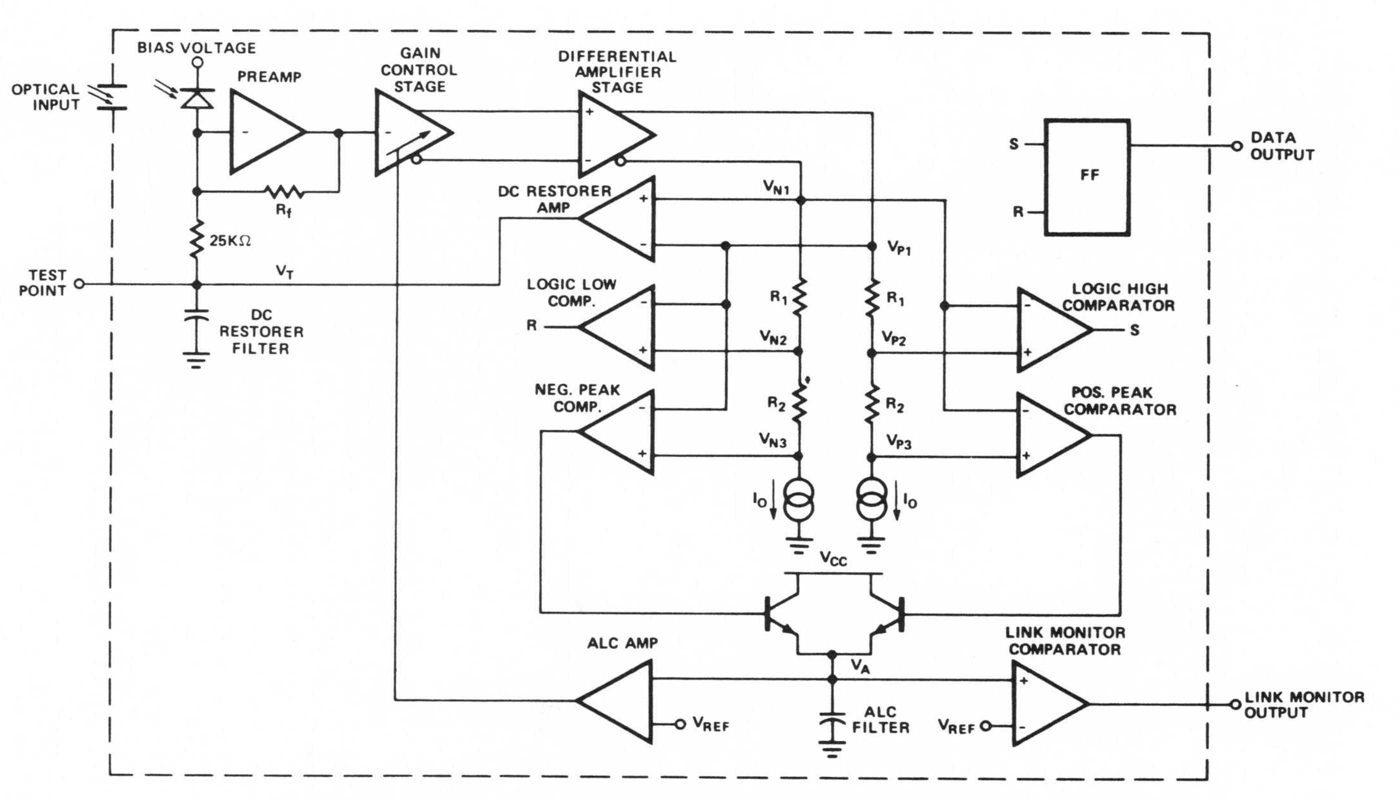

Figure 8-14 Schematic of Data Link Receiver. Courtesy of Hewlett Packard Company.

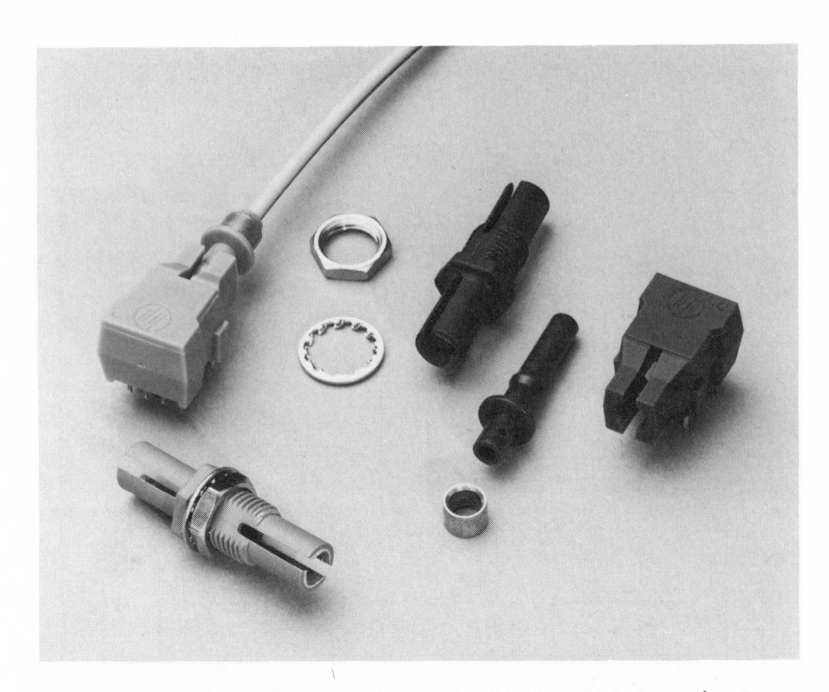

Figure 8-15 Direct Modulation Transmitter (Left, with Fiber Cable Inserted) and Receiver (Right, without Cable Inserted) Modules, Plus Miscellaneous Connector Hardware (Photo). Courtesy of Hewlett Packard Company.

nous waveform stays in the same state (on or off) for a long period of time (no transitions), then the optical transmitter will produce a short duration pulse (refresh pulse) of the same type as the most recent previous pulse. Since positive and negative going transitions of the asynchronous waveform must alternate, the received optical signal is dc balanced. This allows for the receiver to be ac coupled and for the two decision thresholds to be set symmetrically around zero level. The pulses produced by transitions and the refresh pulses are used to set the automatic gain control of the receiver and as a means of link performance monitoring. Figure 8-11 shows a photograph of a transmitter module, using an LED source, which employs this transitional coding technique. The optical connector is visible on the right side of the module. The schematic of the module is shown in Figure 8-12. Figure 8-13 shows a photograph of the corresponding receiver module. The schematic of the receiver module is shown in Figure 8-14. These modules allow for the implementation of a link of up to 1 km length with a data rate of up to 10 MHz.

Figure 8-15 shows a photograph of a very simple transmitter (left, with fiber cable inserted) and receiver (right, without cable inserted) which

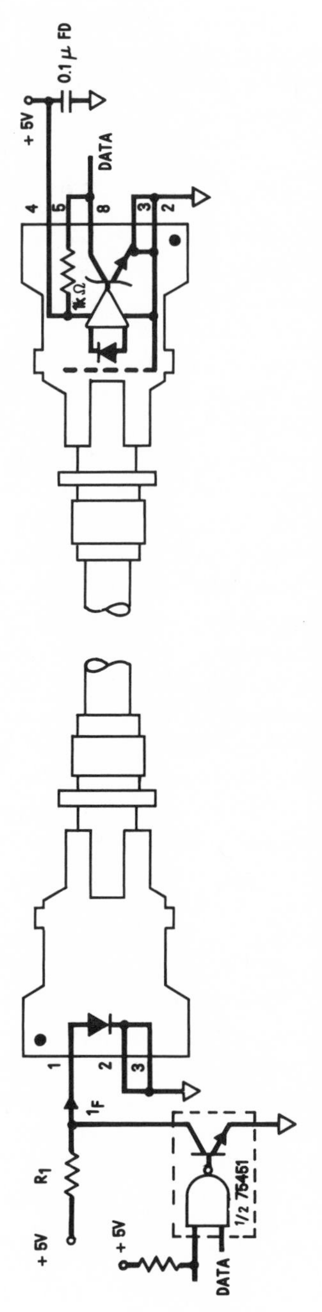

Figure 8-16 Schematics of Direct Modulation Transmitter and Receiver. Courtesy of Hewlett Packard Company.

interface to asynchronous signals by direct modulation. With this transmitter and receiver one can implement a link of up to 5 m in length at data rates up to 10 MHz. Simplified schematics of the transmitter and receiver are shown in Figure 8-16.

8.2 Component Selections for Data Links

As mentioned above, since data links are often short in length, one can make tradeoffs in the allowable loss between the transmitter and the receiver, or in the loss of the fiber, or in the bandwidth capability of the fiber, to accommodate other requirements.

To facilitate interconnections and for coupling more power from LED sources, fibers with relatively large cores (85–200 μm) have been used in data links. The larger core fiber may be somewhat more costly per unit length, and is more susceptible to microbending loss in cabling. However, for short links, this is not necessarily a problem. To allow more light to be coupled from an LED into the fiber, fibers with larger numerical apertures (acceptance angles) have been used. Such fibers tend to have larger amounts of Rayleigh scattering loss per unit length, and are more costly per unit length. In addition, a fiber with a larger NA has less bandwidth for a given length of transmission. Again, for short distances of transmission, this may all be quite acceptable. A fiber with a 0.3 NA can couple 2.25 times as much light from an LED as a fiber with a 0.2 NA. A fiber with a 100-μm core diameter can couple four times as much light from an LED as a fiber with a 50-μm core diameter.

If output power is not critical, one can fabricate extremely reliable LEDs which operate at relatively low current densities (compared to the LEDs used in telecommunications trunking applications), and which allow for other tradeoffs of output vs reliability in the choice of the materials used and the device structure. Output power can also be traded against modulation bandwidth capabilities in the design of an LED for a high data rate, short distance, link. For example, dopants which cause nonradiative recombination, but which reduce the carrier lifetimes can be employed.

One would tend to use electronic circuits which operate from 5 V or −5.2 V (as appropriate) to take advantage of power supplies which are available, even if this results in some compromise in transmitter output power, receiver sensitivity, receiver dynamic range, etc. (depending upon the application).

9

Local Area Networks

The concept of a local area network [26] has emerged along with the revolution in low cost intelligent terminal equipment, distributed computing, and networking of stand-alone personal computers. To a good approximation, the concept of a local area network is as new as the concept of providing connectivity for information transport using optical fibers. This presents us with an interesting challenge. We must consider fiber optics, not as a potential replacement for a well-established set of alternatives for solving a well-defined problem, but rather as one of several alternatives for implementing a new (and sometimes fuzzy) concept.

Even the definition of a local area network is elusive. In Europe, the term local area network often is intended to mean a network for connecting residential and business subscribers to the interexchange telecommunications network. In the United States we would call that network the distribution network or the local access and transport network. In the United States, a local area network is generally an interconnection system within a building or a small geographical area (e.g., a campus) which is accessed by a heterogeneous population of communicating terminals, which is typically privately owned and maintained, and which is generally shared by users who consider themselves part of a common organization (but not always).

A key implication of this is that one might assume (perhaps optimistically) that all of the users are friendly and cooperative (and, in particular, understanding if something goes wrong). This assumption (which remains to be verified by further experience with actual networks) has major impacts on how the network is engineered, in particular, upon how the sharing of resources is controlled (or not controlled).

In this chapter we shall use the definition generally implied in the United States (as given above) for a local area network. We shall discuss distribution networks (public networks for access to the interexchange telecommunications network) later in Chapter 10. In the sections below we shall discuss the following key issues which impact on the architecture of fiber optic local area network concepts:

Application Areas
Networks vs Point-to-Point Links
Measures of Network Performance (Goodness)

We shall then describe a variety of local area network concepts (based on fiber technology) which have been proposed, and comment upon their possible applications, their strengths, and their weaknesses.

9.1 Application Areas

A local area network is an information transport system which allows for the interconnection of a number of terminals which are typically not co-located, and which may often require rearrangeable connectivity. There are a variety of applications for local area networks with significantly different requirements and correspondingly different solutions. That is, a candidate local area network which has characteristics which make it an excellent solution for one application, may be a poor solution for another application. We can list some examples of applications having different requirements. Figure 9-1 provides such a list.

Consider, first, the classical example of the office of the future or the automated factory (being implemented now in some places) where a large number of terminal devices are interconnected in order that they can exchange information and share resources. In the office of the future, these terminal devices are computer terminals, storage devices, centralized computers, data communications equipment (gateways), etc. In the automated factory these terminal devices are programmable controllers for machines, sensors, actuators, central computers, etc. A common characteristic of the office of the future or the automated factory is that a large number (perhaps hundreds or more) of relatively low-speed low-cost terminals must share the network in a way which allows ease of rearrangement (e.g., adding and removing terminals) and ease of use (system is operated and rearranged by nonexperts). In addition, in the office or factory application, the network often covers enough geographical area that the cost of procur-

- OFFICE AND FACTORY NETWORKS
- AVIONICS SUBSYSTEM CONNECTIVITY
- SPECIALIZED COMPUTER SYSTEM/SUBSYSTEM CONNECTIVITY

Figure 9-1 List of Candidate Local Area Network Applications.

ing and installing the transmission medium can be a substantial part of the total cost.

Consider, next, the avionics application. Here one is trying to implement a network to connect subsystems within the relatively small geographical area of an aircraft. There may be a substantial number of subsystems (e.g., radio receivers, signal processors, displays, actuators, etc.) which must be interconnected, but rearrangements involving the installation or removal of terminals are typically much less frequent than in the office application or the factory application. Furthermore, these rearrangements are typically done by highly skilled craft, following procedures carefully documented by engineers. Changes in connectivity associated with different subsystems needing to communicate at different times may occur frequently, but this is built into the networking protocols, and is not observable in detail to the network users. Reliability and predictability are key in this application. Unpredictable delays in obtaining access to shared resources are not acceptable for many of the communications to be carried out. Vulnerability to single point failures or localized damage is a major concern. Many of the communication requirements are for nearly continuous transfer of high clock rate digital signals rather than bursty interactive communications. All of these requirements have a large impact on what might be considered a good candidate network solution. They affect both the hardware aspects and the system (protocol) aspects.

The specialized computer network application is meant to describe a relatively small number (say 10) of high performance computing devices (e.g., mainframes, array processors, storage devices) which must be interconnected in a reliable and rearrangeable manner. It is assumed that the entities being connected are much more expensive than in the office of the future application, and that rearrangements of the network (adding and removing terminals) are infrequent, and carried out by highly skilled persons. The geographical area covered by the network is probably small compared to the office of the future application. What is important in this network is reliability, adequate communications capacity (when needed), predictability for throughput delays and resource availability, ease of physical engineering (e.g., noise immunity), and ease of interface (reasonable protocols). In other words, the users of this network are sophisticated, but they do not want their computing resources to be wasted unnecessarily on the process of interfacing to the network. These users would typically prefer to pay more for readily available communications capacity than to use buffer space and CPU time in order to obtain more efficient sharing of network resources.

In Section 9.4 below we shall describe a variety of significantly different optical fiber local area network alternatives, and we shall point out how they can serve these different applications.

9.2 Networks vs Point-to-Point Links

Many of the technical considerations of point-to-point links are applicable to multipoint to multipoint networks as well. However, there are a few key considerations which have a significant impact upon what can and what cannot be done with fiber optic local area networks. Perhaps the most important of these is the problem of limited available system gain, and its impact on the number of passive access couplers (taps) in a fiber LAN. Available system gain refers to the allowable loss between an optical transmitter and an optical receiver (which can be present while meeting receiver performance requirements). The available system gain must be allocated to transmission loss, connector loss, and margin, in a point-to-point link. In a multipoint-to-multipoint configuration with passive access couplers, the available system gain must also be partially allocated to the losses of these couplers (as will be discussed in Section 9.4 below).

Consider a transmitter—receiver pair operating at a digital rate of 100 Mb/s. Figure 9-2 shows a comparison of the available system gain for optical fiber and copper cable technology. If we assume the use of an LED-based transmitter and a *p-i-n* detector-based receiver, then the available system gain is about 20 dB. If we assume a laser-based transmitter and an APD-based receiver, then the available system gain is about 45 dB. On the other hand, for a metallic cable transmitter with a 1-V output into a 50-Ω coaxial cable, and a receiver limited by thermal noise, the available system gain is about 88 dB.

The large available system gain of the metallic cable system allows for the use of passive taps which remove only a very small amount of power from the cable and/or which couple only a very small amount of power into the cable. The insertion loss of such a tap is typically small.

For the optical systems, the limited available system gain necessitates the use of taps which couple significantly more power into/out of the fiber, resulting in a high insertion loss (due to the power removed and due to reciprocity).

Note that in point-to-point transmission applications, the modest available system gain is not a serious problem, because the loss of the fiber is very low (less than 0.5 dB/km for single mode fiber, for example).

TYPICAL TRANSMITTER OUTPUT		TYPICAL RECEIVER SENSITIVITY AT 100 Mb/s	
LED:	-15 dBm	PIN	-35 dBm
LASER:	0 dBm	APD	-45 dBm
1 VOLT, 50 Ω	13 dBm	"kT" LIMITED	-75 dBm

Figure 9-2 Available System Gain Comparisons.

In addition to the limitations on the design of fiber optic LANs imposed by their modest available system gain, one must also take into account the dynamic range and synchronization problems which frequently (but not always) arise in multipoint to multipoint networks. In an LAN it will often be the case that one of a number of transmitters accessing the network can produce a short message (packet of optical pulses) which must be received and processed by all of the receivers accessing the LAN (even if the packet is only addressed to one receiver, they must all read the address). Each receiver must quickly accommodate its gain (if adjustable) to the level of the incoming packet of optical pulses, and must synchronize its clock (if it uses a clocked regenerator) to the phase of the incoming packet. These factors were important in our discussion of data links in Chapter 8 above, but may be even more important in LANs, due to the presence of multiple transmitters.

9.3 Measures of LAN Performance

The performance of an LAN is often measured by the efficiency of utilization of communications capacity (maximum digital rate which can be supported by the medium). This is generally only one of a number of important measures of performance, and may even be a minor one. This paradox arises because the concept of a multiple access network began with satellite networks, where the cost of the medium (including the satellite) per unit of bandwidth was very high. Later, as local area networks were implemented with metallic cables of limited bandwidth, concern for the efficient use of that bandwidth persisted. However, as one begins to implement a real LAN (particularly a fiber optic LAN) one must also be concerned with such things as efficient utilization of electronics (particularly high speed electronics), ease of interfacing terminals to the network, delay, predictability of the availability of bandwidth to a given terminal, reliability, rearrangeability, etc. All of these factors determine whether the network is useful and cost effective. For example, a very efficient network, which requires high speed terminals accessing the network to have very large buffers, may not be considered a good network by actual users.

The following is a listing of factors of importance in measuring the performance (or goodness) of a local area network. For any given application, different factors may be more important than others.

Network Hardware and Software Cost: How much does it cost to buy and install the hardware, and to buy the software internally used by the network controllers (if any), as a function of the number of accessing terminals and the features provided?

Efficiency of Utilization of the Medium: What is the average amount of information being carried over the fiber links of the LAN, compared to the maximum rate the links can support?

Efficient Utilization of Electronics: Is the complexity of the electronics required to interface the network (including buffers in the accessing terminals) being traded appropriately against other options (simpler protocols, less efficient use of the available fiber bandwidth, etc.)

Delay: Is the total delay from the time a terminal has information to transmit to the time that information is successfully delivered to its destination (and perhaps acknowledged) acceptable?

Delay Uncertainty: Is the uncertainty in the transport delay due to contention for network resources or transmission errors acceptable?

Rearrangeability: How difficult is it to add or remove accessing terminals? Are other terminals' communications disturbed or interrupted in this process?

Number of Terminals: How many accessing terminals can the network support?

Reliability: What is the mean time between failures of a network element? What are the effects of various failures on the network?

Maintainability: How difficult is it to diagnose, locate, and repair failures?

Vulnerability: How can localized damage or improper operation of an accessing terminal affect the network?

Bandwidth Per Terminal: How much bandwidth can each terminal get when needed? Can terminals having different bandwidth requirements be accommodated in a tailored manner?

Broadcasting: Can a given terminal easily send the same message to several (or all) other terminals?

Expandability: As new requirements appear can they be accommodated by techniques such as electronic and wavelength multiplexing?

9.4 Examples of Fiber Optic LAN Designs

Figure 9-3 shows one of the simplest examples of a fiber optic local area network: a passive linear bus. Note that, at the left, transmitting ter-

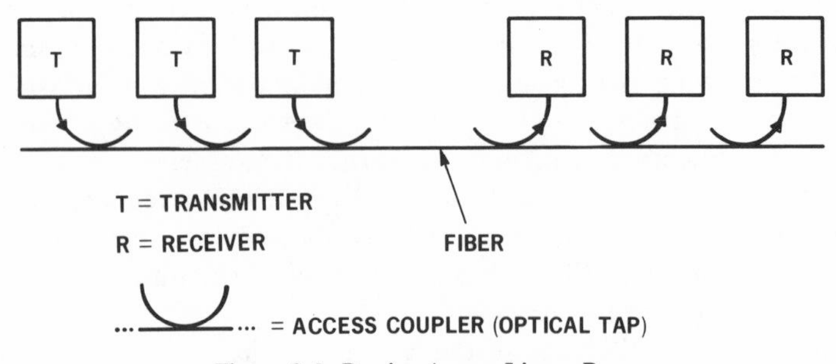

Figure 9-3 Passive Access Linear Bus.

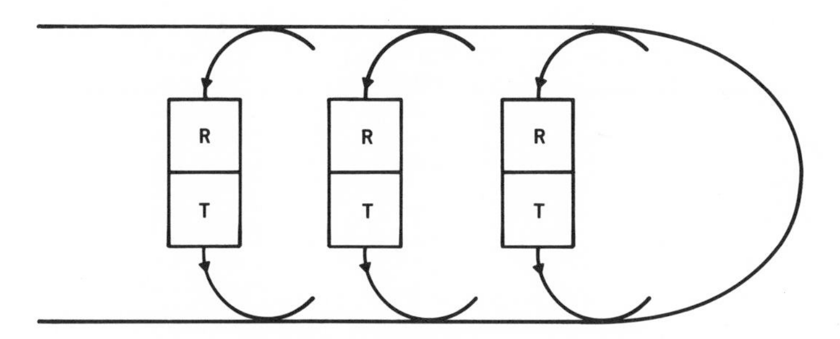

Figure 9-4 Looped-Back Passive Linear Bus.

minals access the bus via passive access couplers (taps), as were described in Section 5.1 above. At the right, receiving terminals access the bus via passive access couplers. Figure 9-4 shows a redrawn version of this LAN where transmitter—receiver pairs are co-located, but where the bus loops around to maintain its unidirectional structure. Each access coupler can be characterized by a loss from the terminal to the bus, which we shall call the access loss of the coupler. Each access coupler can also be characterized by a loss passing left to right (in Figure 9-3), which we shall call the transit loss. For simplicity, we shall assume temporarily that these losses are not sensitive to the spatial field (mode) distribution of the incoming light. We shall return to mode sensitive couplers later in this discussion.

From reciprocity we know that the access loss and the transit loss cannot be specified independently. If the access loss is 3 dB (50% of the light is coupled from the terminal to the fiber or from the fiber to the terminal) then the transit loss must also be 3 dB (plus excess loss to account for light scattered from the tap, and losses of splices required to insert the tap). If the access loss is 10 dB (10% of the light is coupled between the terminal and the bus), then the transit loss must be at least 0.46 dB (10% of the light passing left to right will be coupled out of the bus). For any given access loss, we can calculate the minimum transit loss required by reciprocity (for a tap which is not mode selective). Figure 9-5 gives a listing of access losses from 0 to 15 dB along with their corresponding ideal transit losses (no excess loss). An access loss of 0 dB corresponds to a terminal that accesses the fiber directly without a tap. Also included is a listing of what are called practical transit losses, where 0.5 dB has been added to each of the ideal transit losses to allow for scattering in the tap and splices for inserting the tap in the bus. We shall use this table below in a number of examples, so it is helpful to remember that this choice of 0.5 dB of excess loss is somewhat arbitrary (but reasonable for good couplers and careful splicing techniques).

ACCESS LOSS (dB)	IDEAL TRANSIT LOSS (dB)	PRACTICAL TRANSIT LOSS (dB)
0	∞	∞
1	6.8	7.3
2	4.3	4.8
3	3	3.5
4	2.2	2.7
5	1.65	2.15
6	1.26	1.76
7	.97	1.47
8	.75	1.25
9	.58	1.08
10	.46	.96
11	.36	.86
12	.28	.78
13	.22	.72
14	.18	.68
15	.14	.64

Figure 9-5 Access and Transit Losses for Ideal and Practical Couplers.

Consider the following example. The available system gain (allowable loss between the transmitter and the receiver) between any pair of terminals on the bus is 40 dB. Of this we allocate 30 dB for access losses and transit losses, and we reserve the remaining 10 dB for fiber link losses, margin, connectors, etc. We build the bus with access couplers having 10 dB of access loss and 1 dB of transit loss (see Figure 9-5). The access loss of 10 dB at each transmitter and each receiver uses up 20 dB of the 30 dB allocation for any communicating pair of terminals. This means that we have only 10 dB left for transit losses between any pair of communicating terminals. This, in turn, means that at most 12 terminals can access the bus. (For the assumptions made.)

We can generalize this somewhat by calculating the maximum number of terminals which can access the bus vs a selected value of the access loss (and practical transit loss) from Figure 9-5, provided (for the moment) that all couplers must have the same value of access loss regardless of their position on the bus. Figure 9-6 shows curves of this for three values of allocated total loss for access and transit: 20 dB, 30 dB, and 40 dB. We see for example that with a total loss allocated for access and transit of 30 dB, the maximum number of terminals which can access the bus is 13, obtained with an access loss of 8 or 9 dB. This compares to the number calculated above of 12 terminals obtained with an access loss of 10 dB. With 40 dB of loss budget for access and transit losses, the maximum number of terminals is 22, obtained with an access loss of 9, 10, 11, or 12 dB. With only 20 dB of budget for access and transit losses, the maximum number of accessing terminals is 6.

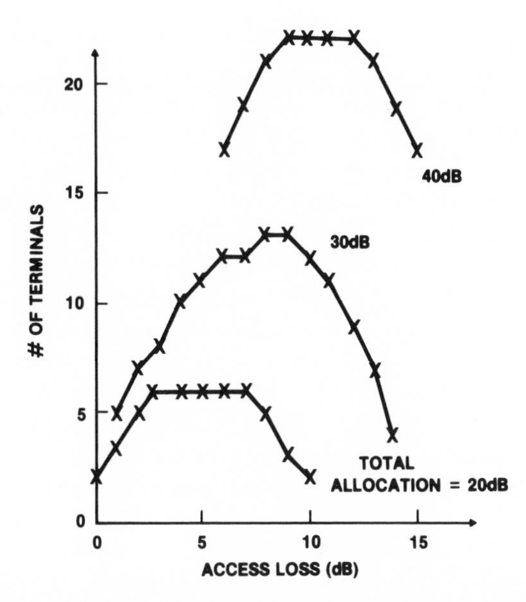

Figure 9-6 Maximum Number of Accessing Terminals vs Access Loss.

We can next ask what improvement could be obtained if all the couplers did not have to have the same access loss. One might argue that as a practical matter it would be difficult to engineer a bus with taps selected according to where on the bus the terminals were (compared to other terminals). However, what is practical depends upon the application, and it is instructive to see what might be gained.

TAP POSITIONS	ACCESS LOSS (dB)	CUMULATIVE TRANSIT LOSS TO CENTER OF BUS (dB)
13, 14	15	0
12, 15	14	0.64
11, 16	13	1.32
10, 17	12	2.04
9, 18	12	2.82
8, 19	11	3.60
7, 20	10	4.46
6, 21	9	5.42
5, 22	8	6.5
4, 23	7	7.75
3, 24	5	9.22
2, 25	3	11.37
1, 26	0	14.87

Figure 9-7 Access and Transit Losses for a Passive Bus Implemented with Tailored Taps.

If we allocate 30 dB to access and transit losses, and if we select taps having integer values of access loss (in decibels) from Figure 9-5, then we obtain the result given in Figure 9-7. In that calculation we assume that half of the terminals are transmitting and half are receiving. We find that there can be up to 26 terminals with access losses having values which are symmetrical relative to the midpoint of the bus. The terminals are numbered consecutively from the left 1–26. The access losses are relatively high near the center of the bus, and decrease to 0 dB (direct access) at the ends. It is also interesting to note that the signal level produced at any receiver by any transmitter is almost identical with this scheme. Thus we see that with the use of access taps tailored to each terminal to match the location of that terminal on the bus we obtain a factor of 2 increase in the maximum number of terminals (in this example) relative to the use of a fixed value for the access loss for all taps on the bus.

We can also ask what improvement might be obtained via the use of mode selective taps. A mode selective tap could provide (for example) low access loss for a particular group of modes in a multimode fiber, while at the same time providing low transit loss averaged over all of the modes of the fiber. This would allow a laser-based transmitter to access the fiber with low loss, while keeping the transit loss (averaged over all modes) relatively low (compared to a tap which is not mode selective). In order for this to work to our benefit, we must assume that the power launched by a given transmitter into a mode selective tap is randomized amongst all of the modes of the fiber before it reaches the next tap. Otherwise the same mode selective mechanism will strongly couple this power out of this next tap, introducing high transit loss for that power (unless we assume that mode selective taps which access different mode groups are used to couple different transmitting terminals).

Whether or not the use of mode selective taps is practical depends upon the application. However, the increase in the number of allowable terminals one obtains is modest (about the same effect as using taps of different values), and furthermore the use of mode selective taps introduces

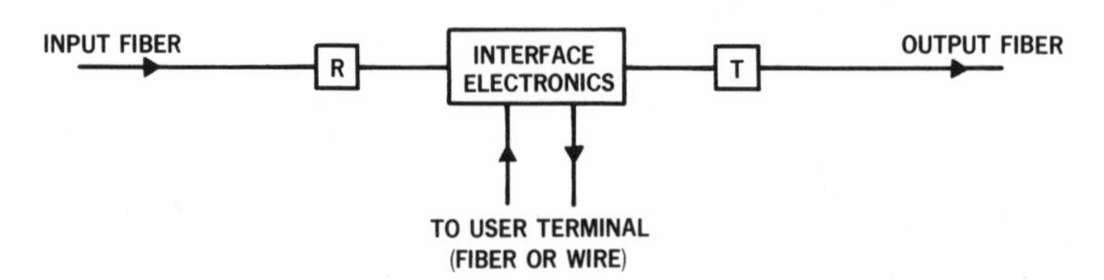

Figure 9-8 Active Access Coupler.

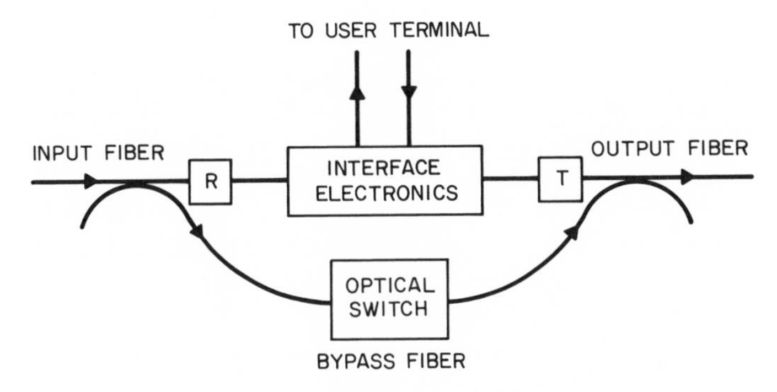

Figure 9-9 Active Access Coupler with an Optical Bypass Switch.

problems associated with modal noise phenomena (as described in Section 6.1).

A passively accessed linear bus (and several of the other busses we shall describe below) is a "party line." The coordination of the use of the bus by multiple transmitting terminals must be considered. We shall do this below in Section 9.5 (Protocols).

Figure 9-8 shows an active access coupler which acts as a repeater. Light from an upstream transmitter terminates on the optical receiver in this tap, to produce an electrical signal. Information is dropped from or added to the electrical signal as required. The modified electrical signal drives an optical transmitter which communicates to the downstream receiver (in the next active access coupler). In this type of bus, every fiber link is point-to-point, and is similar to the type of link discussed in Chapter 8 above. Since the active access couplers (taps) are regenerative (act as repeaters) there is no fundamental limit on the number of accessing terminals imposed by loss budget considerations. It has been often pointed out that an important concern with this type of bus is that the failure of the

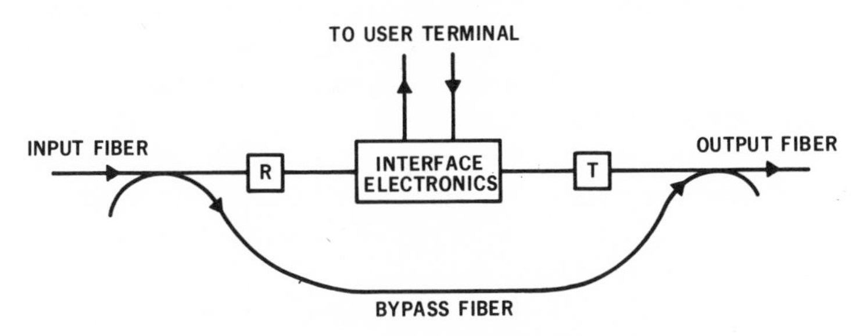

Figure 9-10 Active Access Coupler with Passive Bypass.

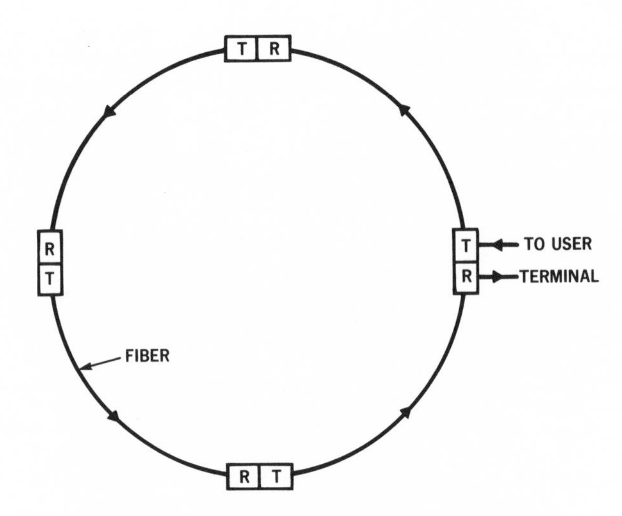

Figure 9-11 Ring-Shaped Active-Access-Coupled Bus.

electronics or optoelectronics in any access coupler will interrupt communication for the whole bus. This problem can be solved with various forms of redundancy. Figure 9-9 illustrates a method using an optical switch of the type described in Section 5.1.4 above. Figure 9-10 illustrates a protection method using a passive bypass approach. Two passive access couplers cause a portion of the light to be coupled out of the incoming fiber and back into the outgoing fiber. The access losses are selected so that under normal circumstances (no failures in the active access coupler) the output power from the transmitter in the active access coupler dominates the bypass signal at the downstream receiver in the next active access coupler. Since the communication is digital, the receiver in this downstream active access coupler can detect the stronger signal without error in the presence of the weak interfering signal. If there is an electronics or opto-electronics failure in the active access coupler it is assumed that the transmitter will not be operating, and therefore the bypass signal will be present alone at the downstream receiver. The downstream receiver presumably has sufficient dynamic range to accommodate this bypass signal, and to detect it without errors.

Figure 9-11 shows a ring-shaped bus which uses active access couplers as described above. In a ring-shaped bus one must be careful that information added to the bus by a terminal does not circulate indefinitely (and thus use up capacity forever).

Figure 9-12 shows a hybrid design where passive couplers are used to access the bus and active repeaters are used to overcome the accumulation

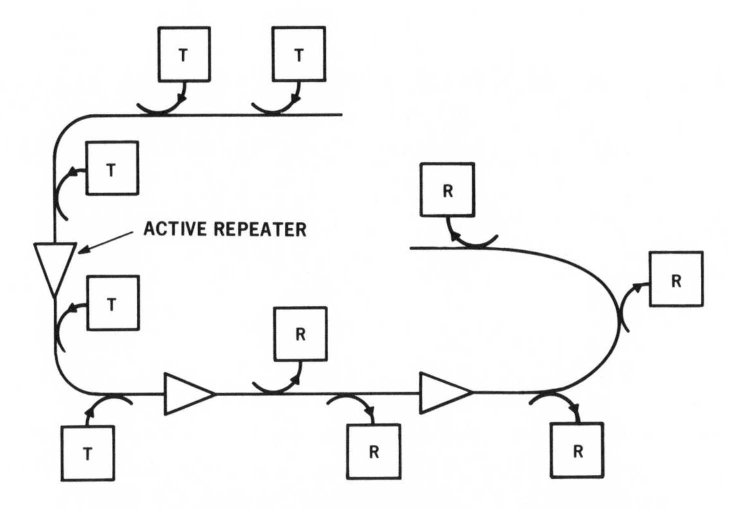

Figure 9-12 Hybrid Active-Repeater/Passive-Access-Coupler Bus.

of access and transit losses. Note that an active repeater also accommodates the varying levels incident upon it from different transmitters and produces a fixed output power level. Thus it not only overcomes loss accumulation, but it also reduces the dynamic range requirements for downstream receivers.

Figure 9-13 shows a very popular passive access optical LAN called a passive star. At the center of the passive star is a star coupler (see Section 5.1.3 above) which divides the power arriving on any incoming fiber approximately uniformly among the outgoing fibers. A practical star coupler will have some excess loss corresponding to the difference between the power arriving on an incoming fiber and the total power distributed to

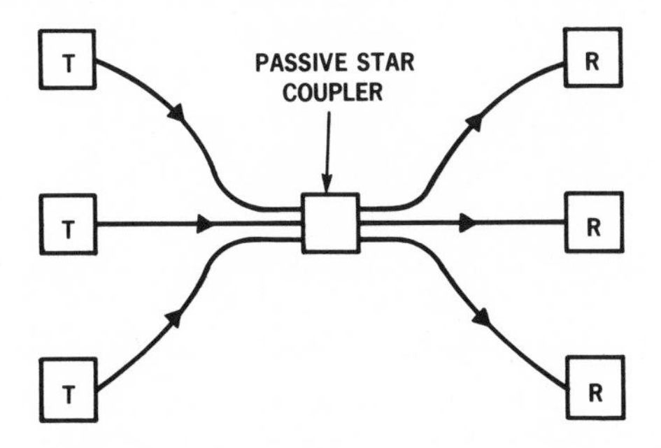

Figure 9-13 Passive Star Coupled Bus.

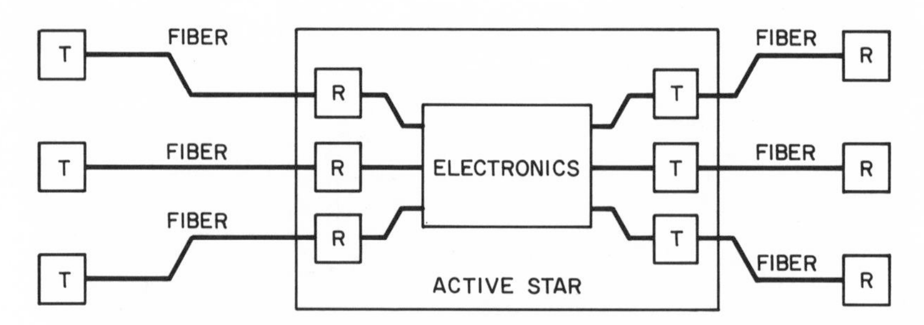

Figure 9-14 Active Star Coupled Bus.

all of the outgoing fibers. A practical star coupler will also have some variation in the power splitting ratio between a given input and a given output. How big the excess loss is and how big the variability is depends upon the number of inputs and outputs, and the quality (cost) of the coupler.

Consider the following example. Suppose that the allocated coupling loss between any incoming fiber and any outgoing fiber is 30 dB. Assume that the star coupler has an excess loss of 3 dB and a variability of input-to-output coupling of ± 3 dB of nominal. This means that the maximum input-to-output coupling loss can be 6 dB worse than an ideal uniform coupler. Since the total loss allocated is 30 dB, then this implies that the splitting ratio of the corresponding ideal coupler with N inputs and N outputs must be only 24 dB (leaving the remaining 6 dB for excess loss and variability of the practical coupler). The input—output loss of an ideal uniform coupler is $-10\log(1/N)$; where N is the number or input or output ports. An ideal uniform coupler with 24 dB of input—output coupling loss would have $10^{2.4}$ input ports and an equal number of output ports. Thus the total number of terminals (equally divided between transmitting and receiving terminals) accessing the bus would be limited to $251 \times 2 = 502$. This compares to the limit of 26 accessing terminals on a linear bus with tailored taps or 13 accessing terminals on a linear bus with equal taps. Another advantage of the passive star network is that the variability between the signal arriving at a given receiver from different transmitters is less than for the linear bus with equal taps. On the other hand, the star shaped network may require more fiber than the linear bus, depending upon the details of where the terminals are located and how the interconnections are routed.

Figure 9-14 shows an active star concept where the central node acts as a repeater for all pairs of accessing terminals. In this configuration the active central node can support a theoretically unlimited number of inputs and outputs, even with a modest available system gain between transmitter

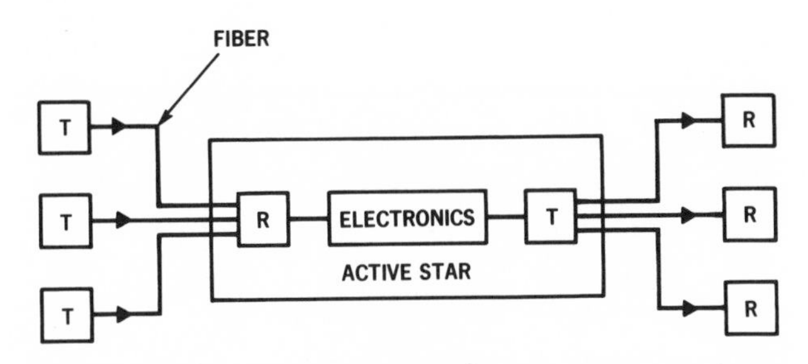

Figure 9-15 Active Star Coupled Bus with Shared Transmitters and Receivers.

and receiver pairs. It is also possible for the incoming fibers to share a common detector, and for the outgoing fibers to share a common source as shown in Figure 9-15. The central active repeater can also serve some protocol functions such as collision detection. Figure 9-16 shows a switched star configuration which allows simultaneous information exchanges between multiple pairs of terminals. The switch could be a circuit switch or a packet switch.

Figure 9-17 shows a concept which follows the allusion of a railroad. That is, one has a locomotive, cars, stations, and trains. The locomotive generator generates a short packet (token) which passes each station (transmitting/receiving terminal pair) in sequence. Each station senses the end of the train and can add a packet (car) if it has one. The train picks up packets from all the stations, and then loops back past all of them again. On the return half of the trip, receivers at each station listen for packets destined for their station. It is not necessary to remove any packets from the train, only to listen to them. When the end of the train is

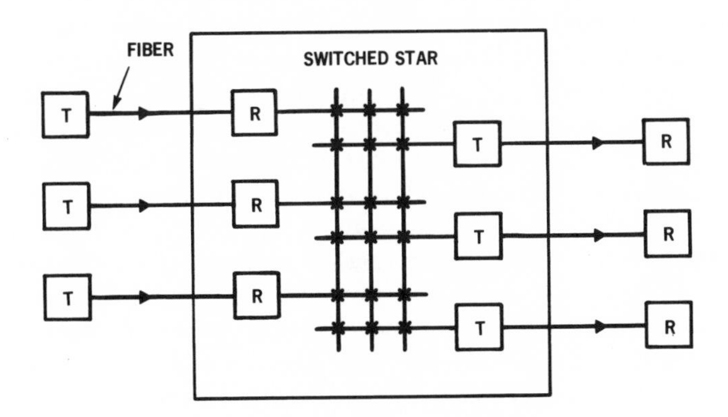

Figure 9-16 Switched Star Configuration.

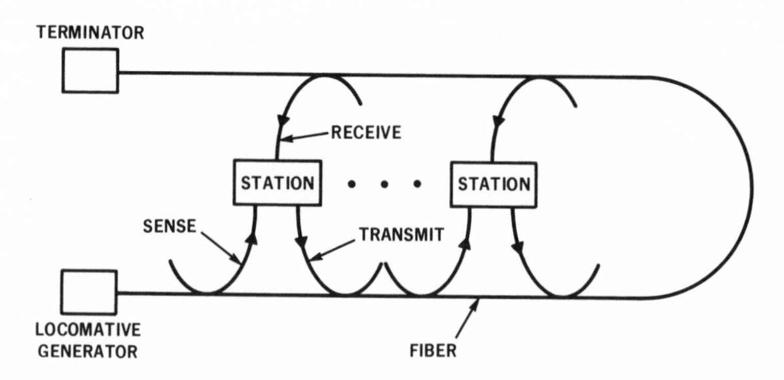

Figure 9-17 "*D*-Net" Configuration.

detected at the terminator, a new locomotive is generated. The advantage of this scheme is that packets from all stations can be transmitted with each pass of the locomotive (token), resulting in higher bandwidth availability for all of the terminals.

9.5 Protocols

A variety of protocols which have been invented for accessing local area networks, in general, can be applied to fiber optic local area networks. We can point out a few considerations which are applicable to LANs in general, but in particular to fiber networks.

A key issue in selecting a network access protocol is the short term averaged bandwidth (data rate) available to the terminals when needed. In principle one can buffer information originating at a terminal to smooth out peaks in bandwidth demand and to collect large packets. However, as data rates increase, buffer storage becomes expensive. Furthermore delays associated with buffering may not be acceptable in interactive communications between terminals (as opposed to one-way file transfers for example). A key parameter which impacts on the available short term averaged data rate is the delay through the network. In some access protocols each opportunity to transmit a packet must be followed by one or more one-way delays through the network to allow the packet to leave the network, or to allow a token to circulate, or to await an acknowledgment (as examples). If one must wait for one or several one-way delays with each packet transmitted, then one is tempted to transmit a packet that is long compared to the one-way delay. Suppose the physical size of the network is 5 km (one-way distance between the most separated transmitter and receiver). The one-way delay is then 5 km × 5 microseconds/km = 25 microseconds. At a data rate of 1 Mb/s this corresponds to only 25 bit intervals. However, at a data rate of 1 Gb/s this corresponds to 25,000 bit intervals. The need to buffer 25,000 bits to efficiently access a 1 Gb/s bus from a 10 Mb/s terminal might prove to be a problem. Even if this is not a problem, it may become one at longer distances and even higher data rates. To solve this problem, protocols which allow multiple packets on the network at the same time are typically utilized.

A popular protocol used in coaxial LANs is the carrier-sense multiple-access with collision detection (CSMA/CD) protocol. Here a terminal listens to the network to see if another terminal is transmitting. If the network appears idle, the terminal can proceed to transmit, but must discontinue transmission if it detects a collision (simultaneous transmission) with another terminal's transmission. The implementation of the collision detection mechanism has proven to be problematic in a number of fiber network designs. In the passively accessed bus or star for example, attempts have been made to detect collisions via the level of the received optical signal. However, due to the variability of the signal level, at a given receiver, obtained from different transmitters, it is difficult to reliably detect a weak signal in the presence of a strong one. One might attempt to detect collisions by listening to one's own transmission and looking for errors. However, again due to the variability of received signal levels, it is possible to correctly receive one's own signal while other receivers are experiencing error causing collisions. It is important to the functioning of the CSMA/CD protocol that one detect collisions rather than using higher levels of the protocol to initiate retransmissions. As the network delay becomes long compared to the duration of a packet, the carrier sense mechanism becomes ineffective in preventing collisions. This can be a problem on physically large networks operating at high data rates with high utilization.

The token passing protocol provides more disciplined access to the network (eliminates collisions unless the protocol is violated accidently), but there is still a long delay between transmission opportunities on geographically large networks unless multiple tokens are allowed.

Protocols which assign bandwidth from a centralized node have the potential for coordinating usage without long delays between opportunities to transmit packets. However, centralized control is often avoided by LAN designers, often for reasons which are more a matter of preference than of technical necessity.

The use of a local area network does not necessarily require packet communication. A popular LAN concept uses a continuous stream of bits which passes by all of the accessing terminals in a ring or a bus shaped network. It is assumed that each access coupler is an active repeater which can modify the bit sequence passing through it. This continuous stream is

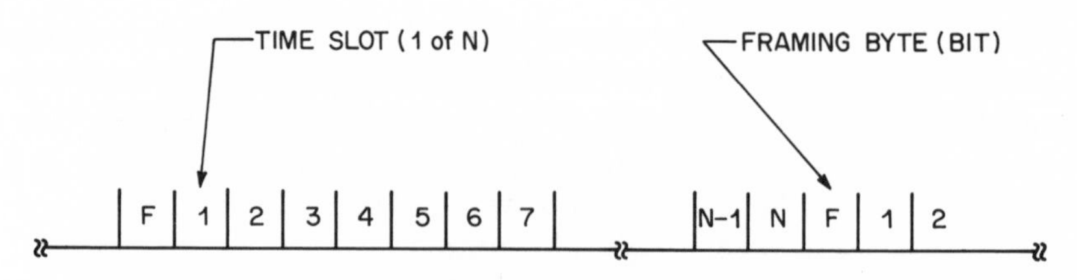

Figure 9-18 Slotted TDMA Frame Structure.

broken up into time slots (channels) which repeat periodically. For example, a 65.6-Mb/s sequence of bits could be divided into 1024 alternating time slots of eight bits each (plus one extra slot used for identifying where one is in the sequence of time slots). Thus every 8200 bit intervals, an eight bit time slot would be available for each of the 1024 separate channels. This is shown in Figure 9-18. The number of bits per second available in each of the 1024 channels is 65.6 Mb/s ÷ 1025 = 64 kb/s. An accessing terminal wishing to communicate with another terminal (or terminals) would obtain permission to use one of the available time slots (or more than one if needed) for the duration of its communicating session. A communicating session could last a millisecond, a minute, or indefinitely, depending upon the application. With this scheme, in the example given, up to 1024 simultaneous information exchanges could take place. In each 64-kb/s channel, the communication can be continuous or packetized. Furthermore, one or more of the 64-kb/s channels can be used with a multiple access packet protocol by several or all of the accessing terminals, while the remaining time slots are circuit switched. For example, one of the 64-kb/s channels could be used for coordinating which terminals obtain access to the other 64-kb/s channels.

Problems

1. In a particular linear bus (Figure 9-4) the allowable loss between any transmitter and any receiver is 25 dB. Assume that the fiber loss is negligible. Assume that all access couplers must be the same, and must be one of the "practical" types listed in Figure 9-5 (e.g., 3 dB access loss and 3.5 dB transit loss). Calculate the maximum number of transmitters and receivers N, vs the access loss in dB (see Figure 9-6 for guidance).

2. In the above, assume instead that each accessing terminal can have a different access coupler chosen from the "practical" couplers in Figure

9-5. Assume an equal number of transmitters and receivers. Find the best values of access loss for each terminal on the bus to maximize the number of terminals (see Figure 9-7 for guidance).

3. An active access coupler (tap) with bypass (see Figure 9-9) operating normally (no failures) can tolerate at its input on interfering optical signal (bypass signal) which is as much as 5% of the desired optical signal. The allowable transmission loss between any optical transmitter and the optical receiver it is talking to cannot exceed 25 dB (total of passive access coupler and fiber losses). What values for the two access couplers (chosen from the "practical" values in Figure 9-5) allow for the maximum fiber loss (maximum transmission distance) in the fiber links between active access couplers, while still allowing reliable communication (bypass) if one terminal fails? (What are the access loss values of the two couplers and what is the maximum link loss for the fiber?) You must select the couplers and the fiber loss so that the bypass signal is below 5% of the desired signal in normal operation and there is enough bypass signal to meet the 25 dB constraint in bypass operation.

4. A practical $N \times N$ star coupler has 3 dB of excess loss and $\pm$ 3 dB of input-to-output coupling variability. If the allowable loss between any input and any output cannot exceed 21 dB, what is the maximum allowed value of N?

10

Analog Links for Video, Telemetry, i.f. and r.f. Remoting

The applications we have discussed so far in Chapters 7–9 above are concerned with the transportation of signals which can assume one of two binary levels, and where the underlying information can be recovered without degradation, even if the times of transition from the lower to the upper level are disturbed slightly in the transmission process. That is, we have been concerned with digital waveforms.

There are applications where it is economically unattractive or technically impractical to convert an analog signal to digital form for transmission over a fiber facility.

In such a situation it may be possible to modulate the optical source in some analog fashion and to recover the information signal by analog demodulation following the optical receiver.

It should be pointed out, however, as a preview to what will follow below, that it is difficult to obtain high quality transmission (low distortion and a high signal-to-noise ratio) using analog techniques with optical fiber systems. This is due to some fundamental limitations and practical limitations which will be described below. On the other hand, today's limitations are often set aside by tomorrow's innovations.

10.1 Analog Modulation Techniques for Use with Optical Fiber Transmission Facilities

A number of standard and not so standard modulation techniques have been used to interface analog signals to optical transmission facilities. The simplest of these is direct intensity modulation. The analog information signal is assumed to have a controlled peak amplitude, to have an average value of zero, and to have a bandwidth B. This waveform $m(t)$ is added to a constant offset so that it is a positive quantity. The resulting positive waveform is used to modulate the power of an optical source as follows:

$$p_s(t) = P_s\,[1 + k_m\, m(t)] \tag{10-1}$$

In equation (10-1) k_m is defined as the modulation index, P_s is the average optical output from the source, and $m(t)$ is assumed to have a peak value (positive or negative) of unity.

Direct modulation can be accomplished, for example, by modulating the current applied to an LED or laser about a fixed bias point.

It will be shown in Section 10.2 below that limitations imposed by linearity and noise in optical transmitters and receivers make it difficult to achieve high quality analog transmission when one uses direct intensity modulation. An alternative technique is to use subcarrier frequency or phase modulation. In this approach, the underlying information signal $m(t)$ modulates the frequency or phase of an intermediate frequency sinusoid. The modulated sinusoidal signal is typically hard limited to produce a frequency or phase modulated square wave. This two-level waveform is then used to modulate the power output of an optical transmitter, very much like a digital waveform. However, the information needed to recover the underlying message $m(t)$ is contained in the zero crossings of the frequency or phase modulated square wave. Disturbances of these zero crossings caused by transmission noise or distortion are converted to degradations (noise and distortion) in the message recovered from the receiver. The advantage of subcarrier frequency (or phase) modulation is that the hard-limited square wave is relatively insensitive to nonlinearities in the optical transmitter or receiver, and with proper selection of the parameters of the modulation process (e.g., using bandwidth expansion) the signal-to-noise ratio obtained upon recovery of the message $m(t)$ can be larger than what is possible with direct intensity modulation. (The signal-to-noise ratio improvement is the classical fm advantage.)

Another popular analog modulation technique, used with fiber transmission facilities, is pulse frequency modulation. In this approach, a modulator produces a sequence of fixed duration electrical pulses at a rate proportional to the amplitude of a message $m(t)$ applied to its input. In some sense, pulse frequency modulation is similar to hard limited fm, except the resulting pulses always return to zero at a fixed interval after turning on. As with hard-limited fm, the sequence of pulses can modulate the output power of an optical transmitter. One interesting property of pulse frequency modulation is that the modulated signal has an average value proportional to the amplitude of the modulating message $m(t)$. Thus $m(t)$ can be recovered by simply low-pass-filtering the pulse frequency modulated waveform. Alternatively, $m(t)$ can be recovered using an fm demodulator.

10.2 Ideal Performance of Analog Links

In this section we shall derive the performance that can be obtained using analog modulation on fiber links if the performance is limited by quantum noise and thermal noise in the receiver, and possibly by nonlinearity in the transmitter or receiver. We shall ignore effects such as intrinsic transmitter noise, modal noise, modal distortion etc. in this discussion, but we shall return to these effects in Section 10.3 below.

Consider an analog link using direct modulation as described in Section 10.1 above, and as shown in Figure 10-1. The power arriving at the receiver is given by

$$p_r(t) = P_r\ [1 + k_m\ m(t)] \tag{10-2}$$

where $m(t)$ is the modulating message having zero average value and unity peak value, k_m is the modulation index, and P_r is the average value of the received optical power.

In response to this received optical signal, the detector produces a photocurrent. If the detector is a *p-i-n* detector, then the photocurrent is given by

$$i_d(t) = \frac{e\,\eta}{hf} P_r\ [1 + k_m\ m(t)] \tag{10-3}$$

Suppose for the moment that $m(t)$ is varying slowly compared to the bandwidth of the amplifier attached to the detector. Assume that the voltage at the amplifier output is related to the current at the amplifier input by a transimpedance Z_R. The voltage at the output of the amplifier is then given by

$$v_{\text{out}}(t) = \frac{Z_R\ e\eta}{hf} P_r\ [1 + k_m\ m(t)] + n_q(t) + n_{th}(t) \tag{10-4}$$

where $n_q(t)$ is shot noise at the amplifier output caused by the current

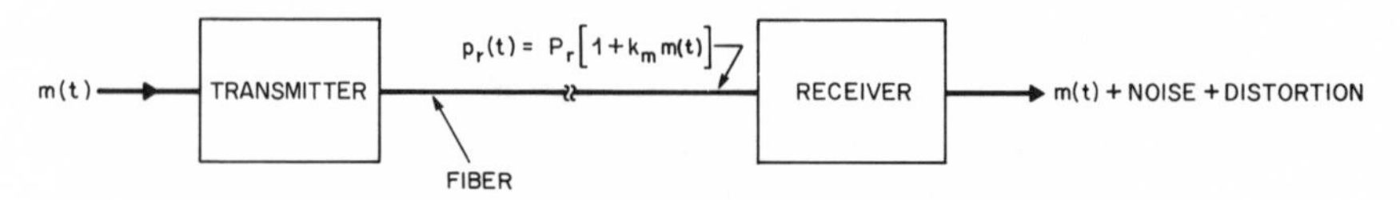

Figure 10-1 Direct Modulation Analog Link.

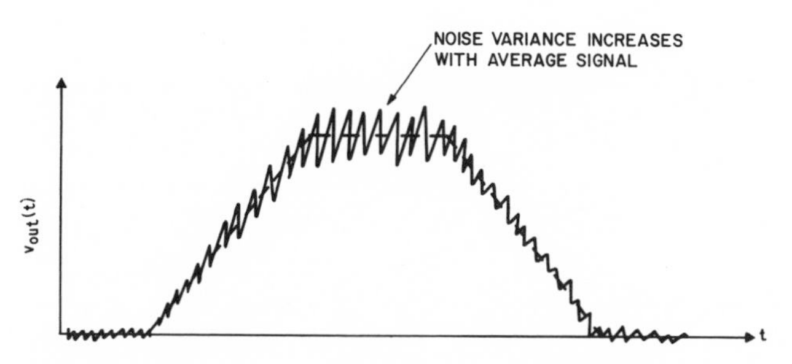

Figure 10-2 Slowly Varying Shot Noise Level at a Receiver Output.

flowing in the detector (a manifestation of quantum noise), and $n_{th}(t)$ is the noise at the amplifier output caused by noise sources within the amplifier itself.

Let us ignore the noise of the amplifier itself, for the moment, and focus our attention on the shot (quantum) noise.

The spectral density of the shot noise at the amplifier input is given by

$$N_{q_{in}}(f) = 2ei_d(t) \tag{10-5}$$

where $i_d(t)$ is the average current flowing in the detector.

Note that as $m(t)$ increases, the average current increases, and therefore so does the shot noise spectral density. [Remember that we are assuming for the moment that $m(t)$ is slowly varying, so it is okay to think of a spectral density which is changing in time.] The noise at the amplifier output has a spectral density given by

$$N_q(f) = 2Z_R^2\, e\, i_d(t) \tag{10-6}$$

If we were to look at the waveform at the amplifier output, with $m(t)$ a slowly varying trapezoidal wave, we might see something like what is shown in Figure 10-2.

In classical communication theory problems we often observe a signal in an additive noise, which is not a function of the signal $[m(t)]$ itself. In that case we define a signal-to-noise ratio by comparing the peak (or rms) signal squared to the noise power in the band of frequencies occupied by the signal.

In the optical case under consideration here, we have observed that the noise power is proportional to the signal amplitude (including its bias level). We can still define what appears to be a classical signal-to-noise

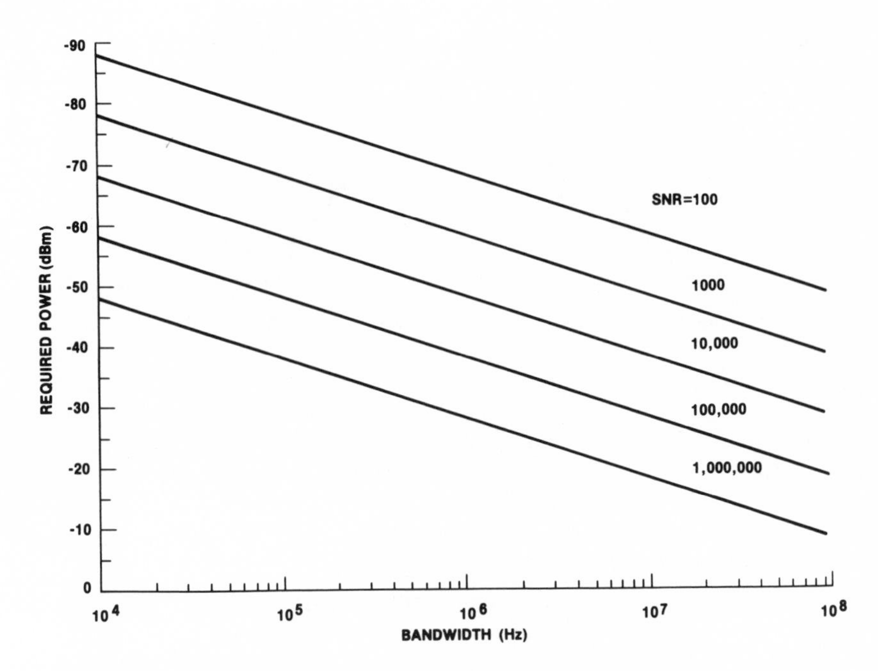

Figure 10-3 Required Received Optical Power vs Bandwidth and SNR (Quantum Limits).

ratio by comparing the peak (or rms) signal squared to the noise power in the band of frequencies occupied by the signal averaged over the values that the signal can assume. Since we assume that $m(t)$ has zero average value, we obtain the averaged noise spectral density from (10-6) as

$$<N_q(f)> = 2Z_R^2 e^2 \eta P_r / hf \tag{10-7}$$

Combining (10-4) with (10-7) we obtain the peak signal-to-averaged noise ratio of

$$\text{SNR} = \frac{\eta P_r k_m^2}{2hfB} \tag{10-8}$$

where we have removed the slowly varying constraint on $m(t)$; where the only noise we are considering, for the moment, is shot (quantum) noise, and where B is the bandwidth passed by the amplifier [and presumably the bandwidth required by the modulation $m(t)$].

This quantum limited signal-to-noise ratio represents an upper bound on the performance which can be achieved by any receiver responding to direct intensity modulated light. It cannot be improved upon by the use of an avalanche detector. In fact, using exotic analysis based on quantum mechanics, one can show that no receiver can provide a signal-to-noise ratio better than that predicted by (10-8) regardless of the mechanisms it employs to extract information from the incoming optical signal.

We can use (10-8) to calculate the minimum required power at the input of an optical receiver to achieve a given peak signal-to-rms noise ratio at the receiver output vs the bandwidth B. This is shown in Figure 10-3, for required signal-to-noise ratios of 20, 30, 40, 50, and 60 dB, assuming $hf = 2 \times 10^{-19}$ J, and assuming a modulation index of 0.5. If we define ηP_r as the detected optical power, then we see for example that for $B = 10$ MHz and an SNR of 50 dB we require a detected power level of -28 dBm (1.6×10^{-3} mW) at the receiver input.

In a practical receiver, amplifier noise will add to the quantum noise to produce a signal-to-noise ratio given by

$$\mathrm{SNR} = \frac{\left(\eta P_r / hf\right)^2 k_m^2}{\left(2\eta P_r B / hf\right) + \left(Z^2 B^2\right)} \tag{10-9}$$

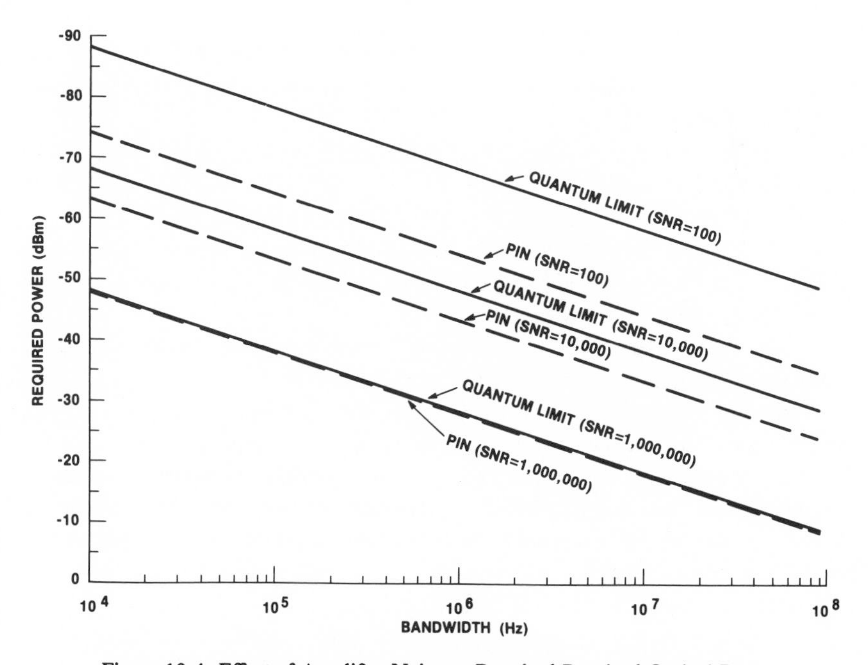

Figure 10-4 Effect of Amplifier Noise on Required Received Optical Power.

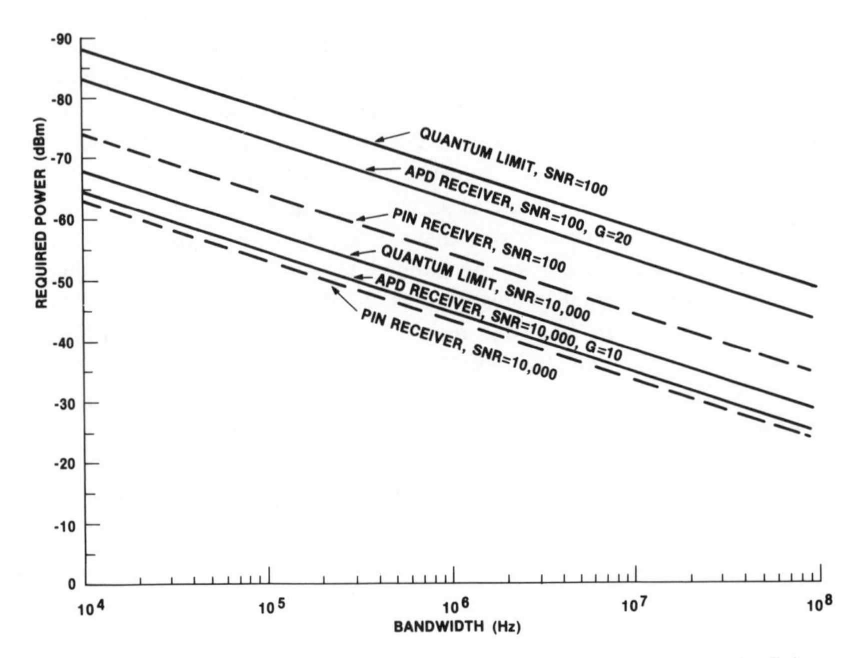

Figure 10-5 Improvement in Required Received Optical Power with Avalanche Gain.

where Z is the amplifier noise parameter (typically around 1000) described in Section 4.2 above.

Figure 10-4 shows effect of amplifier noise by comparing some of the quantum limited curves of Figure 10-3 to curves derived from equation (10-9) with $Z = 1000$, $hf = 2 \times 10^{-19}$ J, and $k_m = 0.5$.

We observe that when the desired signal-to-noise ratio is high (above 50 dB) the performance of the receiver is nearly quantum limited, even in the presence of amplifier noise.

For lower desired signal-to-noise ratios, amplifier noise causes a penalty in the required optical power at the receiver.

If amplifier noise is limiting (e.g., at signal-to-noise ratios lower than 50 dB), one can use avalanche gain to reduce the required optical power (obtain performance closer to the quantum limit). However, avalanche detectors may add nonlinearities which may or may not be acceptable. With the use of an avalanche detector, the signal-to-noise ratio at the receiver output is given by

$$\text{SNR} = \frac{(\eta P_r/hf)^2 k_m^2 G^2}{[2\eta P_r F(G) G^2 B/hf] + (Z^2 B^2)} \tag{10-10}$$

where G is the average APD multiplication (gain) and $F(G)$ is the excess noise factor discussed in Section 4.1.3 above.

Figure 10-5 shows the improvement in required optical power at the receiver that can be obtained with an avalanche detector having an excess noise factor given by

$$F(G) = kG + \left(2 - \frac{1}{G}\right)(1-k) \tag{10-11}$$

with the ionization ratio k chosen to be 0.04. Each curve involving an APD is labeled with the optimal value of avalanche gain G_{optimal}, which is independent of B (for our assumption that the thermal noise in a bandwidth B varies as Z^2B^2).

In addition to the signal-to-noise ratio obtained at the output of the link, one must be concerned with distortions (harmonics) generated by nonlinearities in the link input—output characteristics. Nonlinearities can be caused by the following: nonlinearities in the transmitter current (voltage) input-to-light power output characteristic, nonlinearities in the receiver light power input—voltage output characteristic, nonlinearities caused by modal distortion (discussed in Section 6.3 above), and even by nonlinearities in the fiber power input—power output characteristic (which we have assumed to be negligible and have ignored in this book). If the receiver employs a *p-i-n* detector, and if the electronics in the amplifier are sufficiently linear, then the most important sources of nonlinearity are those of the transmitter and modal distortion.

If the transmitter is biased at some average output level and directly modulated with a sinusoidal current, one will typically observe harmonics in the light output detected by a receiver with a known high degree of linearity. One can measure and plot the relative value of the second harmonic and third harmonic as a function of the modulation index, the bias point, the frequency of the modulation, the type of circuit used to interface to (drive) the laser or LED, etc. To first order, one observes what theory might predict. That is, as one increases the modulation index, the harmonics grow even faster, suggesting that the input—output characteristic is not a straight line. However, the details of these relationships vary considerably from device to device, are highly dependent upon the biasing point, the details of the driver circuitry, and the modulating frequency, and in general do not follow a simple model.

Thus, it is difficult to make any definitive statements as to what the achievable signal-to-distortion ratio is for a given modulation index. This problem is certainly not unique to optical components. It comes up any time real electronic devices are modulated over a range which is a significant portion of their bias level. The equations (10-8), (10-9), and (10-10) above and Figures 10-3, 10-4, and 10-5 suggest that the use of a very small modulation index will typically not provide for an acceptable signal-to-noise ratio in practical systems.

An alternative to the use of direct modulation (at least for laser sources) is to use an external modulator (as described in Section 5.2) above.

The output of an external modulator follows the relationship

$$p_{\text{out}}(t) = P_o \{1 + \sin [k_m m(t) + \theta]\} \tag{10-12}$$

where k_m is the modulation index, $m(t)$ is assumed to have zero average value and unity peak value, and θ is an offset.

The advantage of an external modulator is that it allows more degrees of freedom in the selection and design of the optical source. If the source does not have to be modulated (particularly at high rates), one can use such things as external cavities to reduce its noise, or stabilize its output special characteristics. If the source is not directly modulated, then modal distortion due to chirping of the source spectral content is eliminated.

We can see from 10-12 that the linearity of the external modulator input—output characteristic depends upon the modulation index and the offset in a predictable manner. Figure 10-6 shows calculations of the ratio of second and third harmonics to the fundamental at the output of the modulator [20 log (ratio)] vs. the modulation index k_m for two values of offset: 0 and 0.1 rad (5.7°). One could attempt to improve the linearity somewhat by using predistortion or feedback techniques. The use of predistortion is perhaps a little easier with an external modulator as opposed to the direct modulation case, because the nonlinearity is not only predictable, but it is also dependent on fewer variables. The use of feedback for linearization assumes that the delay around the feedback loop is sufficiently small to accommodate the modulation bandwidth and the amount of linearity improvement being sought.

The use of modulation techniques which expand bandwidth and which are less susceptible to nonlinearities is typically the solution for obtaining even moderate performance on analog fiber links.

For frequency modulation, the signal-to-noise ratio at the output of the demodulator (following the receiver) is $3\beta^3$ as high as the signal-to-noise ratio at the input to the demodulator, where β is the ratio of the peak frequency deviation to the bandwidth of the original message $m(t)$. The signal-to-noise ratio at the input to the demodulator is typically about β times smaller than it would be at the output of a receiver using direct

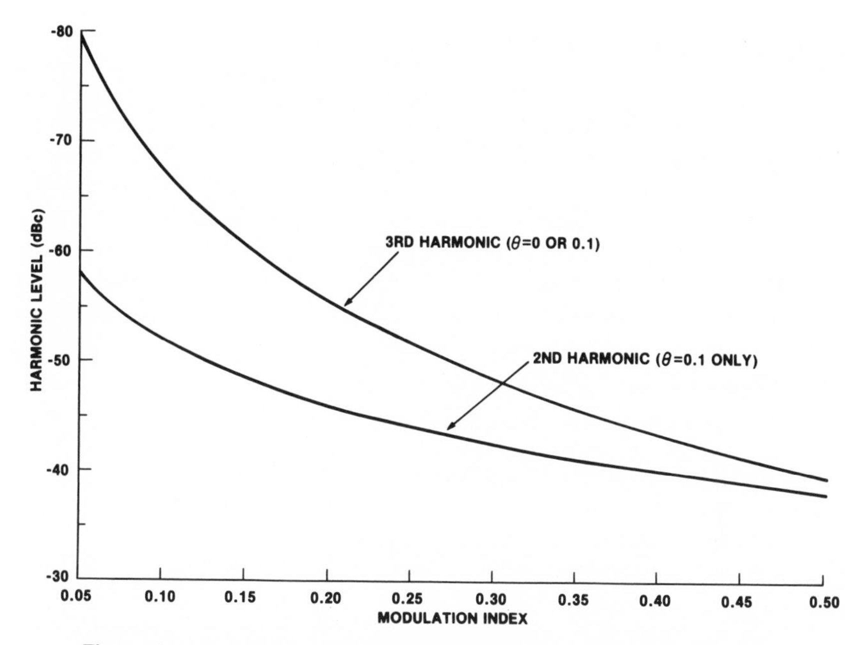

Figure 10-6 Harmonic Levels vs Modulation Index for an External Modulator.

intensity modulation, because the fm signal has a higher bandwidth before demodulation. Thus the net SNR improvement of fm over direct intensity modulation (for a given modulation index) is about $3\beta^2$, or 20 log $(\beta) + 4.7$ in dB. [Note: With an optical receiver, an increase in the bandwidth passed by a factor of β typically results in an increase in the amplifier noise within the band of β^2 or β^3, depending upon the receiver design. However, if quantum noise dominates or if an avalanche detector is being used, the net reduction in the SNR at the receiver output (for a fixed optical power input level and a fixed modulation index) resulting from an increase in bandwidth of β is approximately β^{-1}.] For example, if β is 3, the improvement in SNR is about 14.3 dB. Furthermore, the use of the fm format allows for a higher modulation index, because the fm signal is much less susceptible to nonlinearities. It should be pointed out, however, that a hard limited fm signal is not a digital signal. The information required to recover $m(t)$ is contained in the zero crossings of the frequency modulated waveform. Nonlinearities can disturb these zero crossings under certain conditions, resulting in distortions in the demodulated message $m(t)$.

10.3 Practical Analog Link Applications

The use of analog modulation techniques was first proposed for the transport of television signals. A number of manufacturers currently offer equipment which provides moderate-to-high quality television transport over fibers using subcarrier frequency modulation. Direct intensity modulation has also been proposed for the transport of broadband telemetry (near 0 to several hundred MHz frequency content), i.f. signals (e.g., 70 MHz ± 10 MHz), and r.f. signals (e.g., 5 GHz ± 500 MHz). Some of these broadband signals cannot currently be accommodated by any other method of modulation with today's technology. We shall describe some of these applications in more detail below.

10.3.1 Point-to-Point Television Transport

There are a variety of applications where television signals are transported from point to point, having somewhat different transmission requirements. The most common application is in the distribution of entertainment television programming. Figure 10-7 shows a diagram of a fairly general television distribution network including both vhf/uhf broadcasting (over the air) and cable distribution (CATV) facilities. Television program information originates in studios, and works its way through the network to end users (television receivers). It is not too difficult to show that the quantities of hardware, and the associated cost of equipment, material,

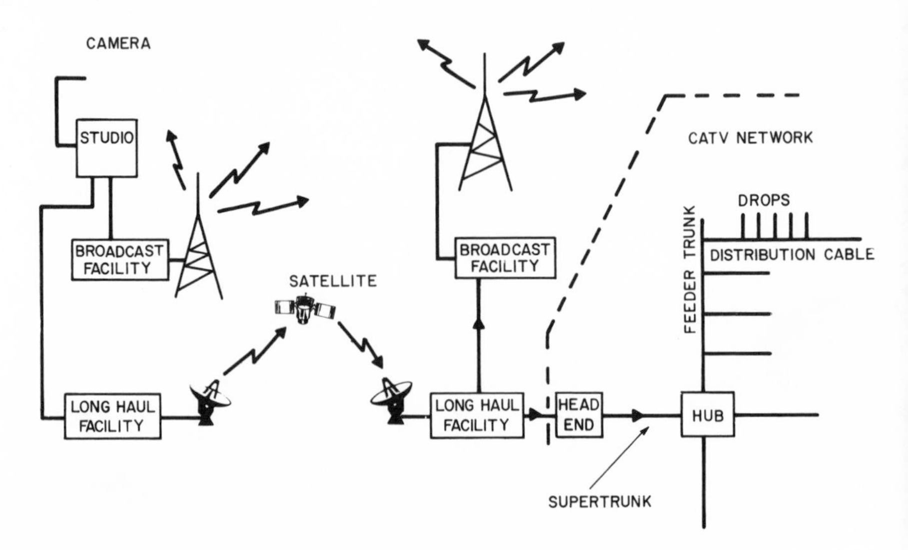

Figure 10-7 Television Distribution Network.

and maintenance, increase as one moves from the studios to the end users. For this reason, transmission performance degradations tend to be allocated more liberally to the portions of the network closer to the end users, and less liberally to the portions of the network which are highly shared. Thus, the transmission performance required of the links between studios and microwave facilities, the point-to-point microwave links (or satellite links), the supertrunks in the CATV network, and any other shared point-to-point links are quite stringent.

Links between studios and broadcasting locations typically carry individual video signals (to be described below). These individual signals are carried for short distances (a few hundred meters or less) over small diameter (e.g., 9mm) coaxial cables. For moderate distances of transmission (e.g., several kilometers), shielded twisted pairs are often used with amplitude modulation techniques, but using very high quality terminal circuitry to maintain low noise and high linearity. For long distances of transmission, the individual video signals are typically used to frequency modulate i.f. carriers, which are then sent over microwave radio or satellite facilities.

In the cable television network, individual program signals are collected at a head end location, and distributed in frequency division multiplexed format (stacked in a frequency band typically between 5 and 400 MHz) over coaxial cables to distribution hubs. The coaxial supertrunks are typically 18–25 mm (3/4–1 inch) in diameter and typically require analog repeaters (amplifiers) every 100–600 m to compensate for the frequency-dependent loss of the cable. The analog repeaters contribute noise and distortion, which accumulates. Thus, the more repeaters used, the more closely spaced they must be in order to control the end-to-end link performance. In fact, if enough repeaters are used, one reaches a point of diminishing returns, where the addition of more repeaters results in less total transmission distance, because the repeater spacings are getting smaller faster than the number of repeaters is increasing! The frequency division multiplexed signals leave the hub on somewhat smaller coaxial feeder cables which also contain amplifiers. Distribution cables split off from the feeders via splitting amplifiers. Individual coaxial cables to end users attach the distribution cables via passive taps.

Economic/technical analyses have been made to determine the most promising applications for fiber optic links in this network. The results of these analyses indicate that the part of the network which extends from the hub to the end users cannot be replaced on a one-for-one basis with fiber. In Chapter 11 of this book we shall discuss a scenario for using fiber in this part of the television distribution network. On the other hand, fiber links have been implemented for some of the point-to-point links between the studios and the broadcasting stations and between the studios and the hubs.

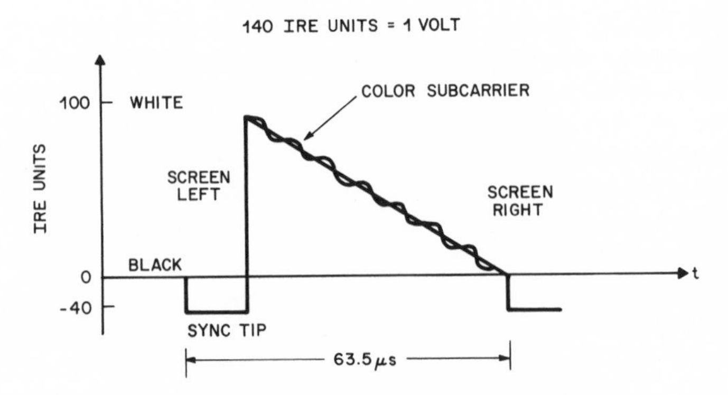

Figure 10-8 NTSC Waveform.

In particular one might use fibers to replace the single video channel links presently implemented over coaxial cable or shielded twisted pairs, the long-distance microwave links, or the multichannel supertrunks in the CATV network.

A U.S. standard (NTSC) analog color television signal occupies a frequency band extending from essentially 0 Hz to 4.5 MHz. Figure 10-8 shows a typical waveform. It has a peak-to-peak amplitude of 1.0 V, including synchronization pulses and picture information. The synchronization pulses occur 15,750 times per second, corresponding to the rate at which the electron beam scans across the screen. The black and white information in the video signal is contained in the level of the signal between synchronization pulses (0—100 IRE units). The color information is contained in the amplitude and phase of a 3.5-MHz subcarrier superimposed upon the black and white information.

The allowable noise, added in transmission by a high quality point-to-point link, is specified by a required signal-to-noise ratio at the link output, typically between 50 and 60 dB for "studio quality" transmission. Distortion is often specified in terms of a maximum allowable "differential gain" and "differential phase." The differential gain and phase refer to a test made with a small amplitude signal at 3.5 MHz (color subcarrier) superimposed upon a slowly varying bias (black and white signal) whose amplitude ranges from the minimum level to the maximum level the composite video signal can assume (0—1 V). The gain and phase, of the end-to-end transmission response of the link to this small high frequency signal, must not vary by more than the specification, as the slowly varying bias transitions through its range of levels. The allowable differential gain is typically between 0.5% and 5%, and the allowable differential phase is typically between 0.5° and 5°, depending upon the application.

The simplest approach one might attempt to implement for transport of a high quality video signal over a fiber is to use direct intensity modulation. When laser sources are used, intrinsic laser noise and modal noise do not allow for the realization of the signal-to-noise ratio requirements of typical applications, although moderate signal-to-noise ratios (35–40 dB) can sometimes be achieved. When LED sources are used, the allowable transmission loss between the transmitter and the receiver is often very small (or negative), as can be seen from Figure 10-3. For example, if we require a peak signal-to-rms noise ratio at the link output of 55 dB, and if we use a modulation index of 0.5, then the required average optical power at the input to the optical receiver is at least −30 dBm (quantum limited performance). An LED can typically launch only −15 dBm into a standard multimode fiber. Furthermore, the linearity of the LED may not be sufficient to achieve the differential gain and phase requirements with a 0.5 modulation index, unless predistortion or feedback linearization is used.

The transmission of several frequency division multiplexed video signals over the same fiber using direct intensity modulation is even more difficult.

For these reasons, high index subcarrier frequency modulation has often been selected as a method of achieving high quality transport of video signals.

Typically a 20–70 MHz intermediate frequency carrier is modulated with between a 5 and 10 MHz frequency deviation by a single video signal. Several frequency modulated i.f. carriers can then be frequency division multiplexed (stacked in different frequency bands) to create a composite signal for modulating the intensity of the laser output. If frequency division multiplexing is used, the locations of the individual fm subcarriers should be selected to minimize the impact of distortion products caused by the nonlinearity of the link.

Studio quality (a somewhat ambiguous term) performance has been claimed for fiber transmission links using subcarrier fm, capable of transporting video signals for several kilometers without a repeater.

It should be pointed out that noise and distortion accumulate as such links are placed in tandem (as in any analog transport facility).

Recently, terminal equipment which converts an analog 4.5-MHz video signal to digital form at 45–135 Mb/s (depending upon the quality required) is becoming available. Using this equipment, one can obtain the benefits of digital transmission, particularly with fiber facilities, where the cost of bandwidth is much less of a factor than with metallic cable and radio facilities. As this video terminal equipment is further integrated, and produced in quantity by multiple vendors, its cost should be equal to or less than the fm modulators presently used.

10.3.2 Broadband Analog Telemetry Links

There are some analog transport problems, involving signals with bandwidths ranging from 100 to 1000 MHz, where adequate quality transmission is difficult to achieve, even with conventional media. These broadband signals cannot be converted to digital form with existing A/D technology, and it is difficult to use bandwidth expansion (e.g., fm) techniques to convert them to more rugged analog formats. The transport alternatives available include baseband transmission over large coaxial cables, direct intensity modulated transmission over optical fiber, and perhaps a few others, depending upon the application (e.g., acoustic surface waves for high-delay–bandwidth–product delay lines).

An interesting example is a delay line with a large bandwidth (e.g., 1 GHz) and a large delay (e.g., 1 ms), but with moderate transmission performance requirements. Such a delay line could be used, for example, for correlating delayed versions of radar returns. A delay bandwidth product of 1 GHz × 1 ms cannot currently be achieved with either coaxial cable or surface acoustic wave approaches. A 1-ms delay corresponds to transmission through 200 km of optical fiber. With single mode fiber losses approaching 0.2 dB/km at 1.55 μm wavelength, it is possible to fabricate an optical cable (or a reel of fiber) of this length with a total loss of 40 dB. If we assume that a laser can launch 0 dBm of average power into the fiber, then the average received power level at the other end of the fiber would be −40 dBm. From equation (10-8) we see that even with quantum limited performance, at 1 GHz modulation bandwidth, the peak signal to rms noise ratio at the receiver output would only be 15 dB (with a modulation index of 0.5).

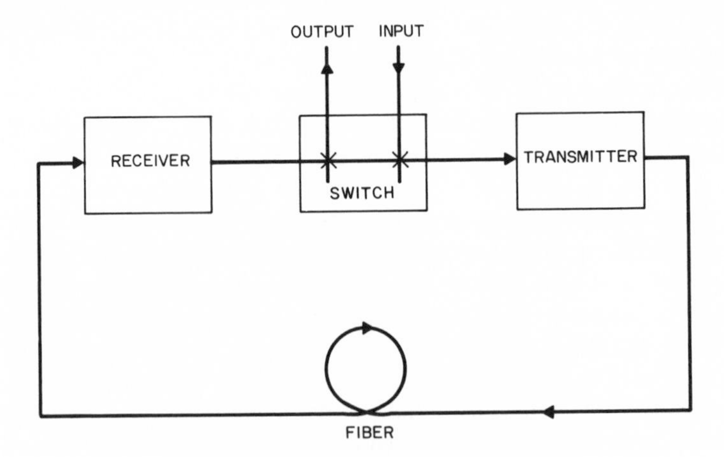

Figure 10-9 Recirculating Delay Line.

In order to obtain better performance, and to reduce the required amount of fiber, one could use a recirculating delay line as shown in Figure 10-9. A broadband pulse, to be delayed, is injected into the delay line and recirculated for N passes though a fiber of length L km, to achieve a total delay of L (km) $\times N \times 0.005$ (ms/km). In each pass through the delay line, the pulse accumulates noise and distortion.

Suppose we assume a launched average optical power level from the optical transmitter (in the analog repeater) into a single mode fiber of -3 dBm, a modulation index (at launch) of 0.5, a net gain around the loop from the input of the transmitter (point A) to the output of the receiver (point B) of -1 dB (for stability), a wavelength of 1.3 μm, a fiber loss of 0.5 dB/km, a *p-i-n* detector with a quantum efficiency of 0.8, and a Z value of 2500. Note that although the modulation index at the initial launch of the pulse is 0.5, this index gets smaller after each pass because of the net loss of 1 dB around the loop.

Figure 10-10 shows the maximum total delay (transmission distance $\times$ 0.005 ms/km) as a function of the number of passes, for a signal-to-noise ratio of 25 dB after passing through the total delay (at the output of the delay line).

We observe that as the number of passes increases, the amount of fiber in the loop (L km) must be decreased in order to maintain a final signal-to-noise ratio of 25 dB after N passes.

Note also the diminishing returns effect, where for more than 10 passes the transmission distance per pass, L, decreases faster than can be compensated by having more passes.

We see that with performance limited by amplifier noise and $Z = 2500$, and with the other system parameter values assumed, a total delay of 0.65 ms is possible.

If we assume that the receiver has a noise parameter, Z, of 2500, and incorporates an APD detector with a quantum efficiency of 0.7 and an ionization ratio of 0.25, we can again calculate the achievable delay as a function of the number of passes, N. This is also shown in Figure 10-10. A delay of nearly 1 ms is achieved for 10 passes around the loop.

Of course other curves could be calculated for different assumptions.

Note that a 25 dB peak signal-to-rms noise ratio at the receiver output is modest enough that intrinsic laser noise might not be limiting.

In addition, the 0.5 modulation index might be compatible with a comparable signal-to-distortion ratio.

Another interesting application is the broadband telemetry link. Suppose we wish to transport a baseband signal having a bandwidth of 200 MHz over a distance of 1 km with a peak signal-to-rms noise ratio at the output of the link of 50 dB.

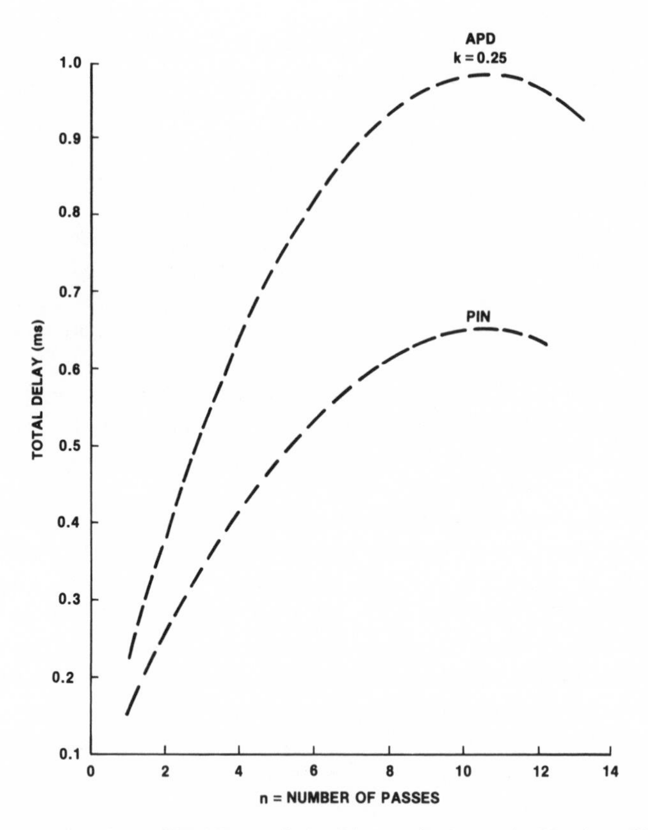

Figure 10-10 Maximum Total Transmission Distance/Delay vs the Number of Passes Around the Loop.

From equation (10-8) we observe that at 1.3 μm wavelength the signal required at the input of a quantum limited receiver (assuming a 0.25 modulation index and a quantum efficiency of 0.7) is −8.5 dBm. From equation (10-9) we can calculate that with a practical receiver having a noise parameter, Z, of 1000, and incorporating a p-i-n detector with a quantum efficiency of 0.7 the required power at the input to the optical receiver is about −8 dBm. Thus one can obtain nearly quantum limited receiver performance without the use of an avalanche photodiode detector. The high required power level at the receiver necessitates the use of a laser source. The short distance of transport implies that transmission loss will not be a major factor (in spite of the high bandwidth of the link). To avoid modal noise, a single mode fiber is appropriate. To avoid mode partition noise and modal distortion it is desirable to use a wavelength where the fiber has zero dispersion (1.3 μm for ordinary single mode fibers).

However, a key practical problem which may prevent an actual implementation from achieving the desired signal-to-noise ratio is the intrinsic laser noise (possibly enhanced by reflections at the launch or from fiber splices). A possible route to solving this problem is the use of a laser with an external cavity or other means for increasing its stability (and hopefully reducing its noise), combined with an external modulator.

10.3.3 Links for i.f. and r.f. Transport

Another interesting application for analog signal transport over optical fibers is shown in Figure 10-11. Here an i.f. or r.f. signal modulates a laser at a center frequency anywhere from a few tens of megahertz (i.f.) to several gigahertz (r.f.). The modulation may be applied directly upon the laser drive current as shown or via an external modulator. The modulated light signal propagates to a receiver over a fiber, where it is detected and demodulated.

An important advantage of fiber in this application is that the performance of the end-to-end link can be virtually independent of the distance of transport for distances between zero meters and a few kilometers (provided that single mode fiber is used). This is because the bandwidth of an ordinary single mode fiber is several hundred GHz-km (neglecting partition noise caused by multifrequency laser operation), and the loss of the fiber can be less than a few decibels over several km of transmission.

This contrasts with metallic media such as coaxial cable or various types of waveguides where the losses per unit length can be substantial, even when large and bulky versions are used. For example, the attenuation of a 9.5 mm (0.375-inch) diameter coaxial cable is about 30 dB/km at a frequency of 500 MHz. The potential for transporting high frequency i.f. and r.f. signals over long lengths of lightweight, nonmetallic, flexible fiber is very attractive.

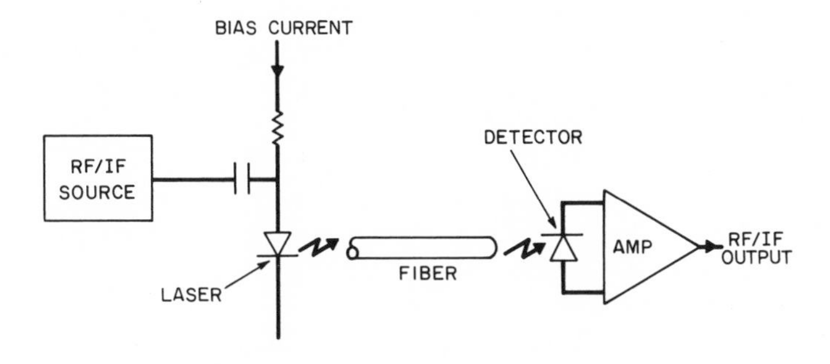

Figure 10-11 Transport of i.f/r.f Signals.

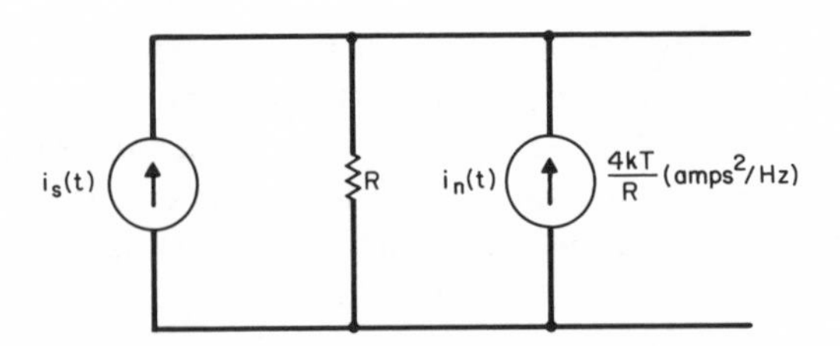

Figure 10-12 r.f. Signal Source Model.

A difficulty in implementing such a link is the ubiquitous problem of practical noise limitations (intrinsic laser noise) and distortion. In some applications, these limitations may not be too serious. For example, a link between a modulator and a transmitting unit in a microwave radio system, where the modulation format is digital (e.g., 16QAM), might be relatively insensitive to noise and distortion. On the other hand, a receiving link in the same system might be very sensitive to noise (which causes errors) and distortion (which causes interference between the desired signal and out-of-band signals). Experimental links have been reported employing direct modulation of long wavelength lasers at frequencies of 5 GHz, with a bandwidth of ±250 MHz. Performance was moderate, as expected, but there is room for improvement in reducing laser excess noise and distortion. Here again, as in the broadband telemetry link, the use of a laser with an external cavity or other stabilization means (to reduce noise) and an external modulator may be a viable approach. Means for improving the linearity of the modulator (e.g., predistortion and/or optical feedforward techniques) would be required with this approach.

In our discussions of analog links we have described the effect of noise added in transmission in terms of a signal-to-noise ratio at the output of the link. In r.f. engineering applications, the noise added in a link is often described in terms of a link noise figure. This measure of performance can be related to our previous results as follows.

Assume that an r.f. signal can be modeled as shown in Figure 10-12 as a current source in parallel with a resistor. The resistor is assumed to have an associated thermal noise of $4kT/R$ (A^2/Hz), corresponding to a noise temperature T and a resistance R.

If this signal modulates the input of an optical link, then the output of the link will be the signal shown in Figure 10-1.

The link has added additional noise, due to shot noise from the detection process, receiver amplifier noise, and possibly other practical noise sources (e.g., intrinsic laser noise).

The noise figure of the link is the ratio of the total noise at the link output in the frequency band of interest to the noise in the same band associated with the signal source noise resistance.

The signal-to-noise ratio at the output of the link is the ratio of the peak output signal squared (excluding noise) to the square of the rms noise at the link output (excluding the noise from the source).

In this definition of signal-to-noise ratio we consider the source noise as zero or part of the signal itself. Thus, we do not include the noise from the source itself in the denominator of the signal-to-noise ratio.

By increasing the modulation index we can increase the peak signal-to-noise ratio, and we can also improve the noise figure. That is, by increasing the modulation index, the noise at the link output from the source noise resistance gets larger, while the noise contributed by the link itself remains fixed.

On the other hand, by increasing the modulation index, we increase the distortion, due to the nonlinearity of the link.

Thus, we can trade noise figure against link distortion by changing the modulation index.

Example: Consider the link shown in Figure 10-13 where an r.f. signal with a 50-Ω equivalent noise resistance and a 273 K noise temperature modulates an external modulator as shown. The amplifier between the signal source and the modulator can be adjusted to set the modulation index. Thus the optical power output of the modulator is given by

$$p_o(t) = P_o \{1 + \sin[\gamma\, m(t)]\} \tag{10-13}$$

where $m(t)$ is the modulating r.f. signal including its associated thermal noise, and γ is a selectible parameter.

We can temporarily approximate this by its linearized equivalent

$$p_o(t) = P_o\, [1 + \gamma\, m(t)] \text{ (watts)} \tag{10-14}$$

The output of the modulator is coupled into a fiber, and transported to a quantum-noise-limited receiver. The attenuation from the transmitter to

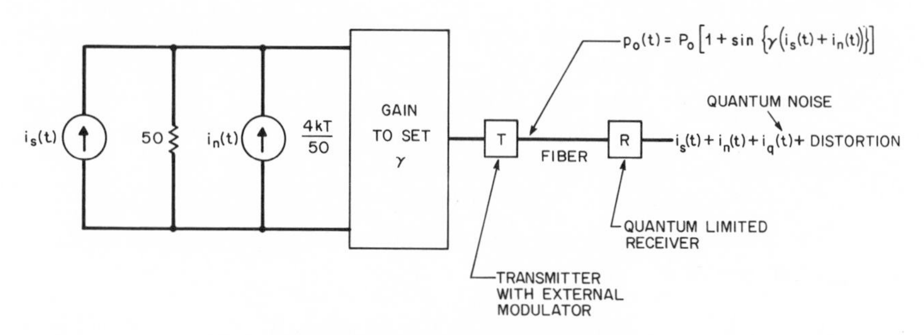

Figure 10-13 r.f. Link Example.

the receiver, including coupling losses and splices, is assumed (arbitrarily) to be 3 dB.

Therefore the signal arriving at the receiver is given by

$$p_r(t) = \frac{1}{2} P_o \,[1 + \gamma\, m(t)] \text{ (watts)} \tag{10-15}$$

The current emitted by the detector (a p-i-n detector with unity quantum efficiency) is given by (approximately)

$$i_d(t) = \frac{1}{2} \frac{e}{hf} P_o \,[1 + \gamma\, m(t)] + n_q(t) \text{ (amperes)} \tag{10-16}$$

where $R = e/hf$ is about 1 A/W at a wavelength of 1.3 μm.

The noise $n_q(t)$ corresponds to the shot noise (quantum noise) of the average current flowing in the detector and has spectral density $eRP_o\ A^2$/Hz.

If we add a current amplifier with gain $A = 2/(RP_o\gamma)$ then the current at the output of the amplifier is given by

$$i_o(t) = m(t) + 2n_q(t)/(RP_o\gamma) \tag{10-17}$$

We see that the link output is the sum of the link input $m(t)$ plus quantum noise from the detection process.

The noise figure of the link is given by

$$\begin{aligned} \text{NF} &= \frac{4kT/50 + 4e/(RP_o\gamma^2)}{4kT/50} \\ &= 1 + 50e/(kTRP_o\gamma^2) \end{aligned} \tag{10-18}$$

We can adjust this noise figure by selecting various values for γ. However, as we increase γ we increase the distortion associated with the nonlinearity of the external modulator.

In r.f. engineering, one often refers to the second- or third-order intercept point of a nonlinear link. This is the value of a sinusoidal link input signal which results in a second or third harmonic equal to the fundamental. In actuality, this is calculated by extrapolation, based on the harmonics that result at lower input levels.

We can calculate the third-order intercept point in our example.

The relationship between the modulator output power and the modulating input signal is

$$p_o(t) = P_o \,\{1 + \sin[\gamma\, m(t)]\} \tag{10-19}$$

For moderate distortion, this can be approximated by

$$p_o(t) = P_o \{1 + \gamma\, m(t) - \frac{[\gamma\, m(t)]^3}{6}\} \tag{10-20}$$

If $m(t)$ is a sinusoid, of amplitude S, then the ratio of third harmonic to fundamental at the modulator output is given by $\gamma^2 S^2/24$.

Extrapolating, we find that the value of S which produces a third harmonic equal to the fundamental is $\sqrt{24}/\gamma$ (amperes).

We can express this as follows. The link r.f. input power corresponding to the third-order intercept point (IP_3) is $50 \times 12/\gamma^2$ W or $57.8 - 20 \log [\gamma]$ dBm.

Combining this result with the noise figure result we obtain the following:

$$\mathrm{NF} = 1 + 2 \times 10^3/(P_o\, \gamma^2)$$

$$\mathrm{IP}_3 = 57.8 - 20 \log (\gamma)\ \mathrm{dBm} \tag{10-21}$$

where P_o is in watts. We see that we can trade the noise figure against the third-order intercept power (referred to the link input) by varying γ. Since we assumed quantum limited link performance, there is no way to improve upon the relationships given in (10-21) except to increase the received optical power level, P_o, or to improve the linearity of the modulator.

11

Broadband Networks

A broadband network (also called a broadband integrated services network) is one which can distribute broadband services, such as video, to end users on demand. The concept of a broadband network using a star shaped topology, and based on fibers for transport, originated in the mid-1970s.

Figure 11-1 shows a listing of services which one might wish to distribute to end users with an advanced technology communications network. Also shown in Figure 11-1 are the bit rates required per channel for each of these services in digitized format. For example, speech is presently transported in digital format using 64 kb/s per speech channel. With advanced speech coding technology, recently emerging, one could transmit higher quality speech or speech at lower data rates, but 100 kb/s should be more

SERVICES	REQUIRED DIGITAL RATE (Mb/s)
ONE-WAY VIDEO (VIEWING)	10-100/CHANNEL
TWO-WAY VIDEO (PICTUREPHONE®)	1-100/CHANNEL
HIGH FIDELITY AUDIO	< 1.0/CHANNEL
DATA	< 0.1/TERMINAL
VOICE	< 0.1/LINE
TELEMETRY	< 0.0001/CHANNEL

Figure 11-1 Services Provided by a Broadband Network, and Associated Digital Rates per Channel.

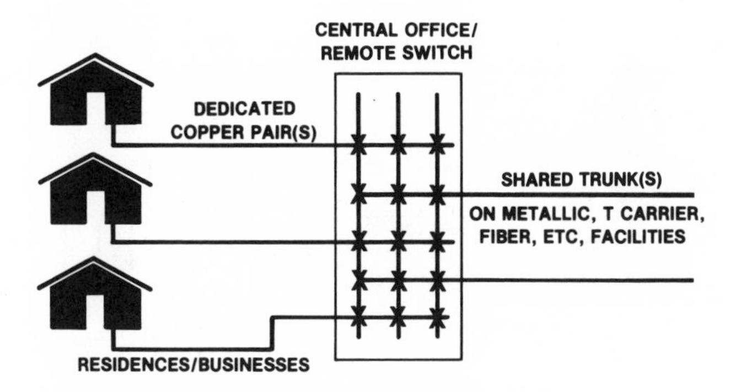

Figure 11-2 Switched Telephone Network.

than enough for any application. Typical present day data transport services provide between 300 baud and 56 kbaud of point-to-point information carrying capability. It would appear that for interactive information exchanges between terminals and information sources, 100 kb/s would be quite adequate (even for browsing through data bases). File transfers could require much higher rates. Telemetry for alarms and meter reading requires just a few bits per second of data exchange. A high quality stereo audio program requires about 1 Mb/s, corresponding to a sampling rate for each of the two channels of 30 kHz, and about 16 bits of information per sample. What really takes up bandwidth is digitized video. A full motion video program must be sampled at about 10.5 MHz (three times the color subcarrier frequency) with about 8 bits per sample. This yields a composite rate of nearly 90 Mb/s.

With complex coding techniques one can remove some of the inherent redundancy of video to reduce the required rate with some tradeoff of quality. For videoconferencing applications, rates below 1.5 Mb/s can be achieved by these types of coders, but with considerable loss of quality when the picture contains significant motion. Extended quality video, an emerging technology, may require data rates of 135—450 Mb/s, unless complex coders are used.

With the present networks in the United States, one can provide switched voice, switched data, broadcast video, and broadcast audio services. One can provide voice, data, and telemetry using the switched telephone network, as shown in Figure 11-2. This network was originally designed to provide switched analog 3 KHz voice channels, but can be augmented to provide data services at moderate bit rates (below 100 kb/s approximately). The copper wire plant is not capable, in general, of supporting high fidelity audio or full-motion video services (either in analog or

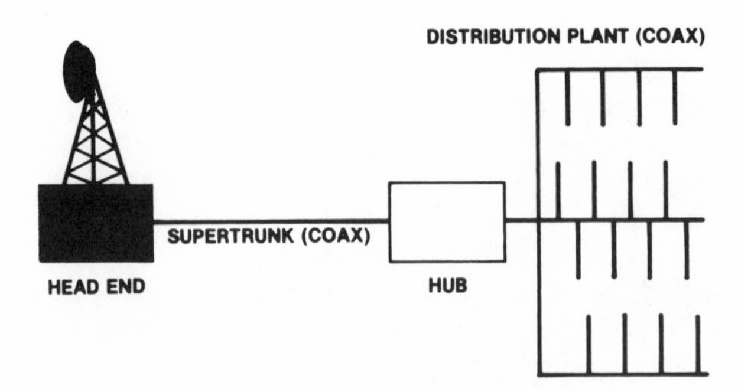

Figure 11-3 CATV Network.

digital form). The CATV distribution network, shown in Figure 11-3, can distribute a finite number of video channels in analog form with moderate received quality. It cannot provide any significant amount of individualized programming, because a limited number of available video bandwidth channels must serve a much larger number of customers. That is, the CATV network is essentially a broadcast network, as opposed to a switched star network (like the existing telephone network).

Figure 11-4 shows a concept which resembles the switched telephone network, but which uses fibers (rather than wire pairs) for distribution.

Each customer (business or residence) has a fiber (or several fibers) interconnecting his or her premises and a remote switching unit.

In the remote switching unit one has an ordinary voice/data switch handling low-to-moderate rate information and signaling (call setup) information. In addition, each remote switching unit has a broadband switching matrix for video, high fidelity audio, and other high bandwidth signals.

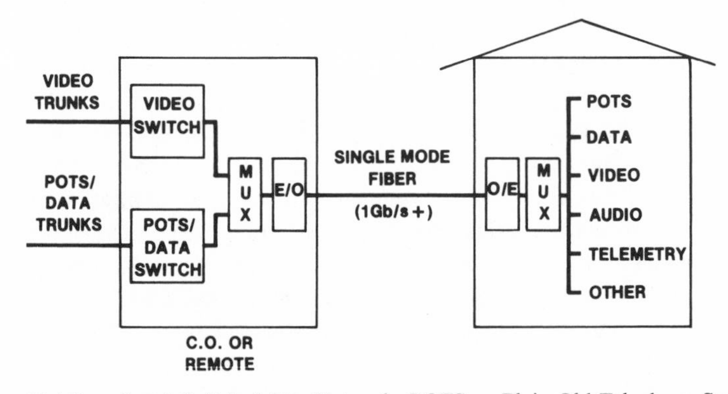

Figure 11-4 Broadband Switched Star Network. POTS = Plain Old Telephone Service.

Information arrives at the remote switching unit via trunks from information vendors and other, more distant, end users.

Several channels of broadband and voice/data/signalling information are multiplexed together at the remote switching unit and delivered to the customer over a fiber (or fibers). At the customer's premises these are separated into individual information services terminating on appropriate customer-premises equipment.

As an example of how this network might function, consider a customer wishing to watch a particular movie at a particular time (e.g., right now). The customer sits at a terminal and dials up the remote unit to request a listing of program availability and cost (this is a data information exchange so far). The remote unit connects the customer (via the voice/data/signaling switch) to a data base which provides the requested information. The customer determines that the particular movie he wishes to see is available, and will cost a specific amount of money to watch. The customer types into his terminal that he wishes to watch the movie at that price, starting immediately. This information (data) is processed, and causes a connection to be set up (via the video switch) between the vendor of the movie and the customer's premises, using one of the broadband channels being transported over the customer's fiber. The vendor of the movie sends the movie signal to the remote switching unit over one of the video trunks. In this way, that particular customer can watch any available movie (or other video information) when he wants to, independent of what other customers are watching. The fiber link to the customer would typically support several simultaneous video services plus other services as well.

11.1 Technologies for Implementation of Broadband Networks

There are a number of technologies which could be used to implement a broadband network, depending upon the services to be provided and the time frame for implementation.

In the near term, the lack of widely available low cost video A/D and D/A converters and high rate digital switching matrices tends to favor the use of analog techniques for handling the video transport. That is, one or two video signals of moderate quality might be distributed to end users over individual fibers using subcarrier frequency modulation or pulse frequency modulation techniques (see Chapter 10 above). In the longer run (e.g., beginning around 1990) it is likely that the appropriate digital coding and switching technology will be economically available for implementation of all-digital broadband networks. For example, if high quality video is to be supported, one might require digital switching matrices capable of switching 135+ Mb/s serial streams. This is a challenging task, but within reach with present and forecasted device technology. The advantages of an all digital network are the well-known ease of maintenance and high quality of service obtainable with rugged digital signals.

The links between the remote switching units and the end users will likely be single mode fibers carrying upwards of 1 Gb/s (several multiplexed digitized video channels, plus voice and data). Two-way transmission over the same fiber may be utilized, if this is more cost effective than using two fibers. Some novel approaches to implementing the customer-to-remote switching unit link may be utilized because of the relatively short length (and low loss) of this link. For example, one might use a laser source at the remote switching unit end of the link to transmit the high bandwidth video signals. However, at the customer's end one might use an LED (even with a single mode fiber) to send the relatively low bandwidth voice, data, and signalling up to the remote switching unit (if the customer has no video sources on his premises). It is conceivable that one might send an unmodulated light signal from the remote switching unit to the customer and have the customer modulate it remotely, and reflect it back.

11.2 Tree Shaped Fiber Video Distribution Networks

The conventional CATV network is a tree shaped network where a number of video signals (as many as 60 with present technology) are broadcast to end users over coaxial cables, as shown in Figure 11-3. One might ask whether there is an opportunity to use fibers in place of the coaxial cables as a one-for-one CATV network replacement, rather than implementing the switched star network described above. This would be exceedingly difficult to do because of the difficulty of transporting many frequency division multiplexed (or even many time division multiplexed) video signals simultaneously over a single fiber, as described in Chapter 10. Fibers are best suited for a switched star configuration where only the channels requested are transported to the end user over his dedicated fiber. One might transmit a large number of digitized video signals per fiber over the trunking part of the switched star network (using very high data rates and wavelength multiplexing techniques), but this is not feasible on the nonshared portions of the network. In addition, the coaxial tree network employs a large number of splitting amplifiers and passive taps. The accumulation of noise and distortion at optical fiber network equivalents of splitting amplifiers would only increase the problems associated with multichannel analog transport over fiber. Furthermore, as was described in Chapter 9 above, passive access is very difficult to implement in fiber net-

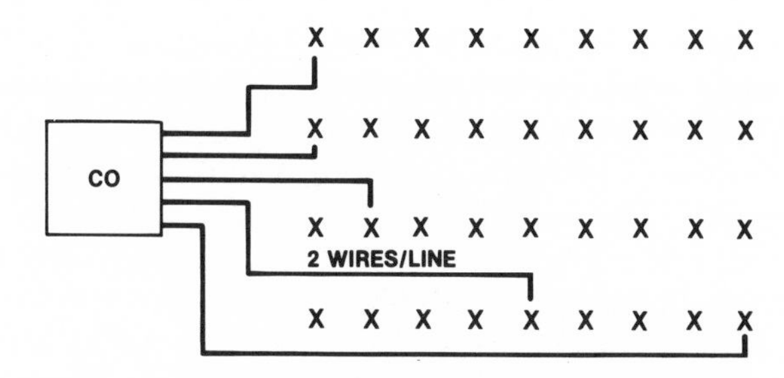

Figure 11-5 Wire-Based Telephone Distribution Network.

works (even more so with analog modulation) because of the small allowable loss between any transmitter and any receiver.

11.3 Loop Carrier Systems

The implementation of broadband networks will occur over a long period of time, starting with demonstration systems to test the services and the technology.

In the nearer term, fiber optic loop carrier systems have proven very effective in bringing fiber technology closer to the end users and in laying some of the groundwork for future broadband networks.

Figure 11-5 shows a diagram of a conventional telephone distribution network, where each user accesses a switching center via a pair of wires. In recent years (beginning around 1970) subscriber carrier systems have been deployed to reduce the amount of copper in the loop plant and to provide an improved grade of service to customers who are far from their central office (switching location). Figure 11-6 shows the concept of a subscriber carrier system. An unattended (remote) electronic terminal is placed near a cluster of end users. Telephone lines (wire pairs) from these end users terminate on this nearby remote subscriber carrier terminal. The

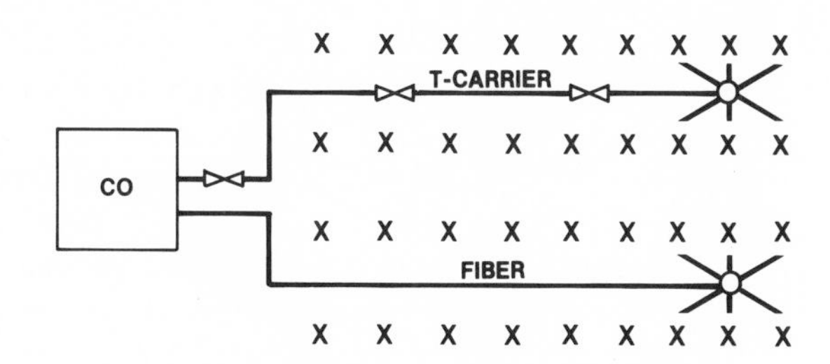

Figure 11-6 Subscriber Carrier System Concept.

connection between the remote subscriber carrier terminal and a complementary terminal in the central office is via a digital carrier system (e.g., T-carrier or fiber). The remote and central office terminals perform functions similar to the trunk carrier terminals described in Chapter 7 above. However, there are some nontrivial call processing and maintenance functions which must be performed as well. For example, when the central office attempts to ring a subscriber's line it applies a 20 Hz, 100 V, ringing signal to a pair of wires terminating on the central office subscriber carrier terminal. The central office terminal detects this high voltage signal, and sends a digital message to the remote terminal telling it to ring the appropriate subscriber's line. The remote terminal has its own 20-Hz, 100 V ringing generator, which it applies to the appropriate pair leading to the subscriber to be rung. When the subscriber goes "off hook" he draws current from his pair. This current is detected by the remote terminal, which sends a message to the central office terminal telling it to draw current from the corresponding pair connected to the switching machine. The switching machine then cuts the connection through. Thus, call setup messages are transported in facsimile fashion, detected at one end of the subscriber carrier system, and recreated at the other end. The same is true for transferring dial pulses (if used), commands to activate coin telephones, special ringing signals for party lines, etc.

Means must also be provided for testing customers lines for continuity, foreign potentials, excessive leakage to ground, etc., without requiring a visit by maintainence personnel to the remote terminal.

Since the central office-to-remote terminal connection is digital, the quality of service delivered to the end users is governed primarily by the lengths of their wire pair connections to the remote terminal, and by the quality of the electronics in the remote terminal and the central office terminal.

When fibers are used to implement the link between the remote terminal and the central office terminal, one reaps the additional benefit of

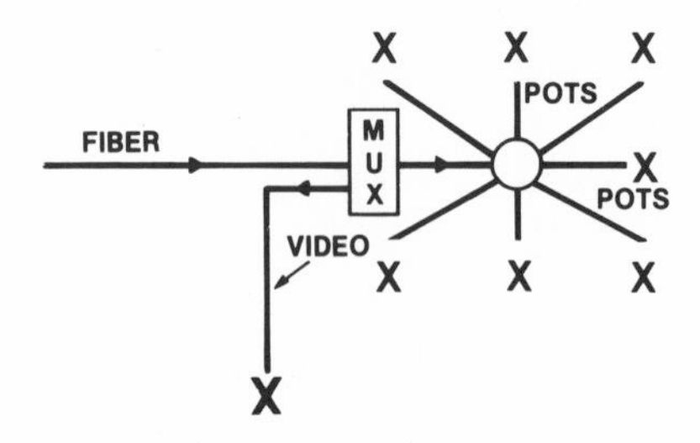

Figure 11-7 Using Excess Bandwidth in Fiber Links to Remote Terminals.

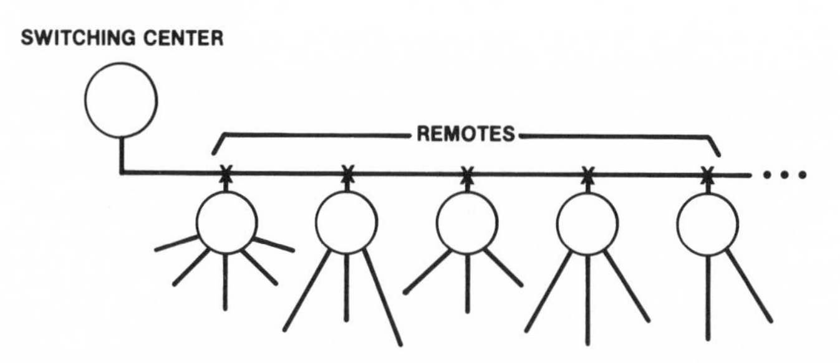

Figure 11-8 Sharing the Capacity of a High Bandwidth Fiber Transmission Facility.

having substantial information carrying capability available for other uses. Figure 11-7 shows an example where a multiplexer (electronic or wavelength) is used to piggyback a high bandwidth service, for a customer in the cluster, on top of the digital carrier link serving the remote terminal.

Figure 11-8 shows a possible configuration where many remote terminals share the capacity of a fiber optic transmission facility and access the amount of data rate required to support the services terminating on them. Some of these remote terminals might be providing only voice and moderate rate data services, while others might provide higher rate data and video services.

If we add switching capability to these remote units, we begin to see the emergence of a broadband switched network.

12

Sensing Systems

The purpose of a sensing system [22, 24] is to acquire information about an environmental parameter of interest (e.g., pressure, voltage, current) and to transport that information to a convenient location for processing. One could build a sensing system where the acquisition function is performed by conventional technology (e.g., a piezoelectric pressure sensor) and the transport is accomplished via an optical fiber link. In this section we shall discuss sensor systems where both the acquisition and the transport are accomplished by optical or optoelectronic means.

12.1 Sensing Mechanisms

Optical sensing systems can be divided into two classes which work on considerably different principles, and which have substantially different areas of application. Incoherent sensing systems are those which employ a sensor which has a variable transmission loss or reflection loss, which is read out by illuminating the sensor with an optical signal. Coherent sensing systems operate by dividing an optical signal into two parts which travel different optical paths, and which are then allowed to combine and interfere. One of the paths contains a sensor which modulates the phase of the optical signal passing through it. The other path is a reference path.

12.1.1 Incoherent Sensing Mechanisms

Figure 12-1 shows the simplest type of incoherent sensor system, where an environmental parameter modulates the insertion loss of the sensor. An optical transmitter produces a probe signal, which is detected by an optical receiver. An example of such a sensor would be a length of semiconductor material whose band-gap energy is nearly the same as the photon energy of the probe signal. Such a sensor would tend to absorb light passing through it by an amount which would be influenced by its

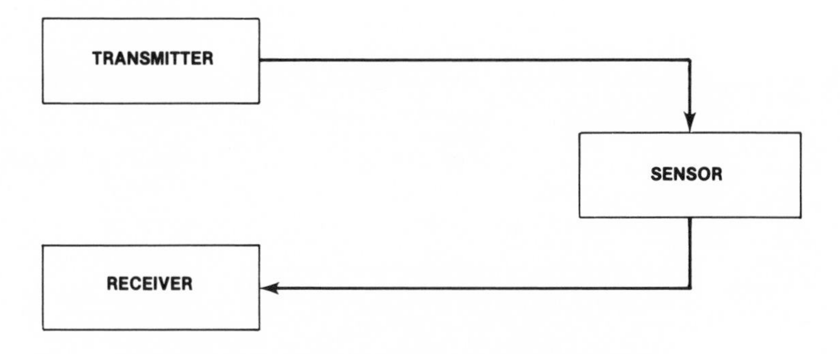

Figure 12-1 Incoherent Insertion Loss Sensor System.

temperature (which modulates the band gap). Thus a change in the temperature of the sensor could be detected by variations in the received signal level. It has been proposed that two probe signals could be launched through such a sensor. One probe would have a photon energy below the band gap, and thus would not be absorbed by the sensor. This probe could be used as a reference to track out variations or uncertainties in the losses of the fibers and connectors in the system.

Figure 12-2 shows a reflection sensor system, where the sensor reflects back a portion of the light launched by the probe transmitter. An example of such a reflector might be a liquid crystal whose reflectivity is temperature sensitive. An even simpler example might be the reflection from the end of a cleaved (flat) fiber which is either above or below the level of a liquid in a container.

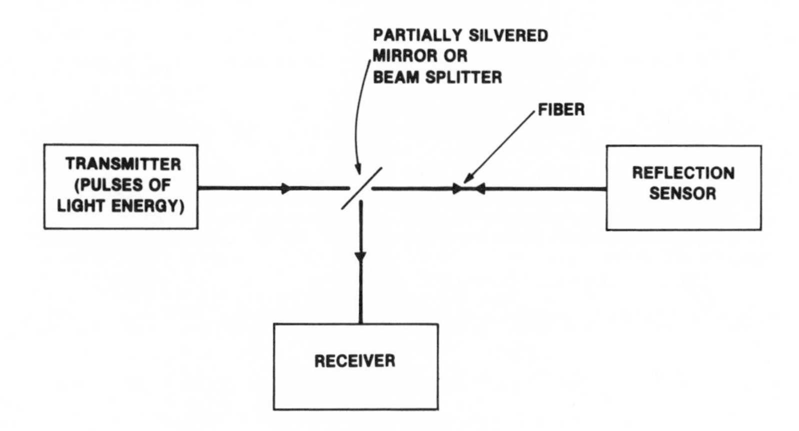

Figure 12-2 Incoherent Reflection Sensor System.

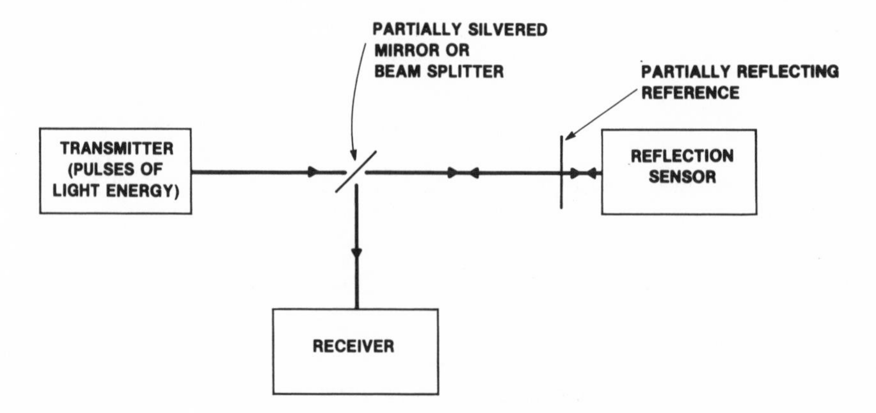

Figure 12-3 Incoherent Reflection Sensor System with Reference Reflector.

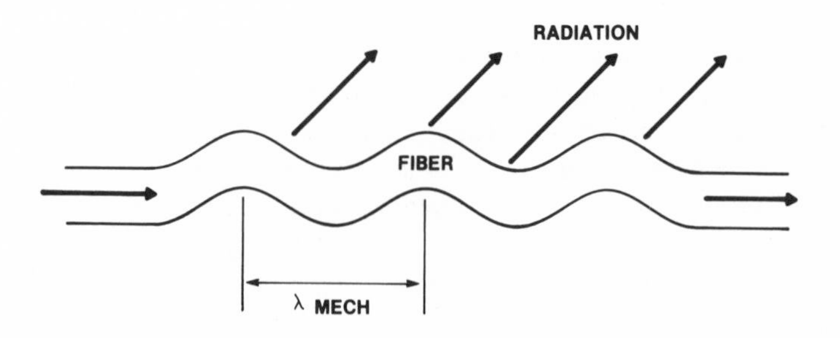

Figure 12-4 Microbending as a Mechanism for an Insertion Loss Sensor.

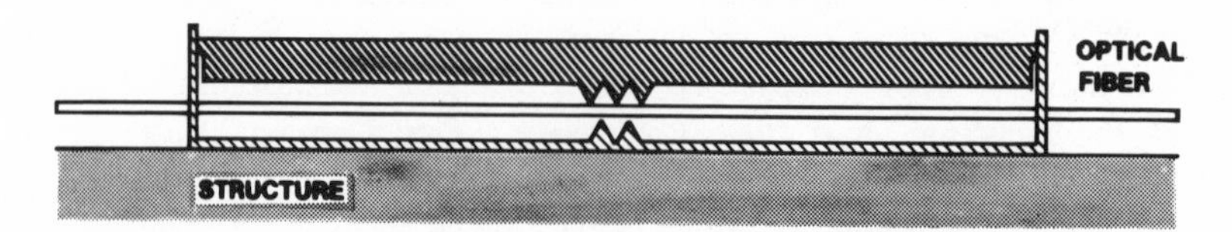

Figure 12-5 Microbend Sensor Schematic. (Courtesy of TRW, Inc.)

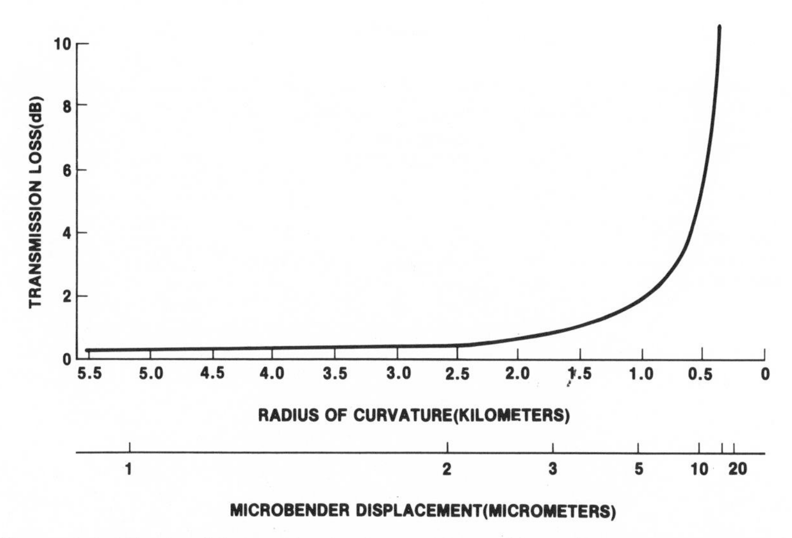

Figure 12-6 Microbend Sensor Insertion Loss vs Displacement. (Courtesy of TRW, Inc.)

Figure 12-3 shows a modification of the reflection sensor system where a partially reflecting reference reflector is placed in front of the variably reflecting sensor. The reference reflector is used to track out uncertainties or variations in the losses of the fiber or connectors in the optical path, and also to track out changes in the output of the probe transmitter.

Figure 12-4 shows how the mechanism of microbending can be used to implement a variable insertion loss. In the design of cables, great effort is expended to avoid bends in the fiber with periods of millimeters to centimeters. Such bends can cause coupling of guided modes to unguided radiation modes. In a microbend sensor, microbending loss is used to advantage to detect mechanical movements of minute amplitude (μm). Figure 12-5 shows a schematic of a microbend structural-bending sensor, where a small displacement of the sensor's relatively flexible lower bar is caused by the bending of a structure to which the sensor is attached. Figure 12-6 shows a curve of measured sensor insertion loss vs radius of curvature of the structure, and the corresponding displacement of the sensor jaws. We see that a 5μm jaw displacement results in a 1 dB increase in insertion loss of the sensor.

12.1.2 Coherent Sensing Mechanisms

Figure 12-7 shows the basic principle of a coherent sensing system. An optical signal is divided into two parts by a beam splitter (power

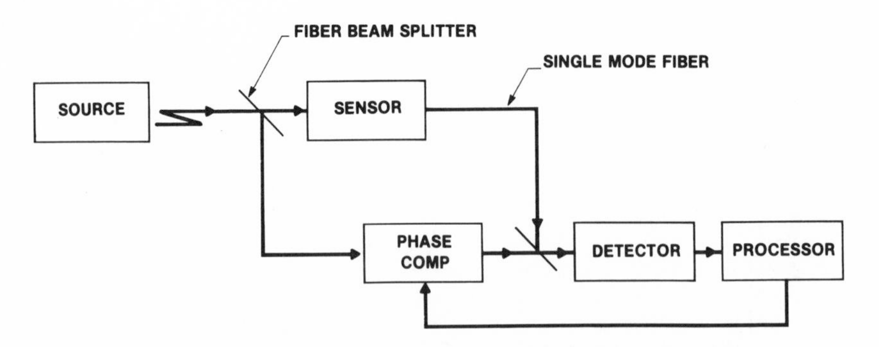

Figure 12-7 Basic Coherent Sensor System.

divider) as shown. One part travels through a reference path, and the other part travels through a sensing path. The signal in the reference path is added to the signal in the sensing path by another beam splitter, and the combined signal falls upon an optical detector. The optical detector responds to the total optical power falling upon it, which consists of the power from each of the paths, plus cross terms due to the coherence of the two signals.

For example, suppose the output of the transmitter is (ideally) an optical sinousoid at optical frequency f.

The two optical signals arriving at the detector from the two paths (in the same spatial field pattern and polarization) have amplitudes A_1 and A_2, and phases θ_1 and θ_2. The combined signal complex field falling on the detector is given by

$$\tilde{E}_{\text{Total}} = [A_1 \exp i(\theta_1 + ft) + A_2 \exp i(\theta_2 + ft)]\Psi(x,y) \quad (12\text{-}1)$$

where $\Psi(x,y)$ is the field pattern focused on the detector. The detector responds to the integral of the magnitude squared of this field over the detector's sensitive area. Thus the current produced in the detector is given by

$$i_d(t) = \frac{\eta e}{hf} I_o \left[A_1^2 + A_2^2 + 2 A_1 A_2 \cos(\theta_1 - \theta_2)\right] \quad (12\text{-}2)$$

where

$$I_o = \int |\Psi(x,y)|^2 \, dx \, dy$$

We see that if θ_2 is given by $\theta_1 + \pi/2 + \Delta$ rad, then the current produced in the detector will be given by

$$i_d(t) = \frac{\eta e}{hf} I_o \, (A_1^2 + A_2^2 + 2 A_1 A_2 \sin \Delta) \qquad (12\text{-}3)$$

Thus, if the two paths are in approximate phase quadrature, a small disturbance of the phase delay in either path will produce an approximately linear change in the current produced by the detector.

Since a meter of propagation distance corresponds to 10^6 wavelengths (at 1 μm wavelength), a small disturbance of the propagation characteristics of a relatively short sensor can cause a noticeable change in the detector output.

12.2 Examples of Sensing Systems

In Section 12.1 we discussed a temperature sensor based on the insertion loss of a semiconductor with a temperature sensitive band-gap (Figure 12-1), and another based on the variable reflectivity of a temperature sensitive liquid crystal (Figure 12-2). These same temperature sensitivities could be harnessed for measuring other parameters. For example, a tem-

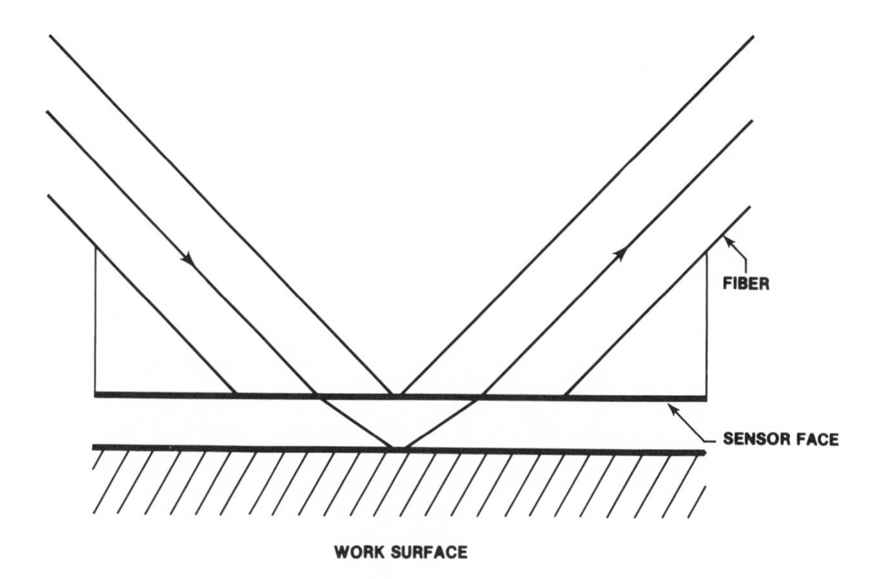

Figure 12-8 Distance Sensor Using Reflected Light.

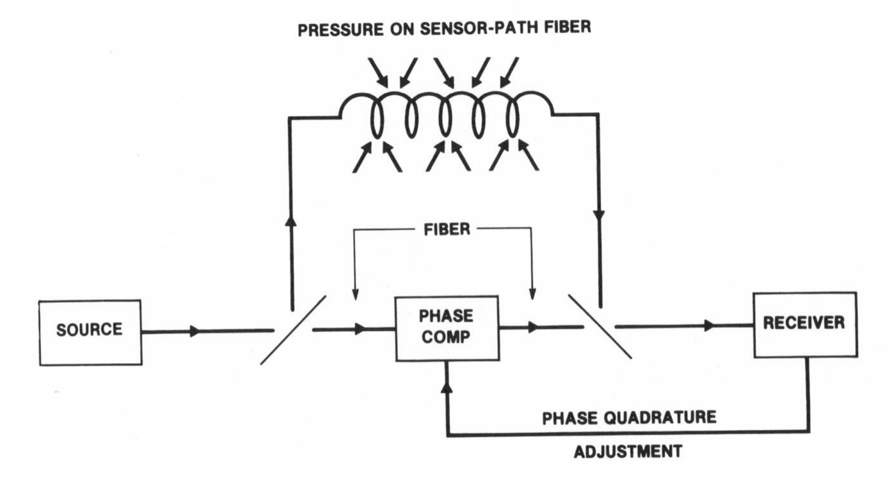

Figure 12-9 Hydrophone.

perature sensor could be coated with a material which absorbs microwave energy, thus converting this energy to a temperature rise.

Figure 12-8 shows an example of an insertion loss sensor which actually works on a reflection principle. The amount of light coupled from the upstream fiber to the downstream fiber depends upon the distance of the reflecting work surface from the fiber ends.

Figure 12-9 shows how a coherent sensor can be used as a hydrophone. Pressure in the medium around the hydrophone sensor is converted to strain on the sensing fiber, which in turn modulates the phase delay through the sensing fiber. Some significant challenges in making a sensitive and dependable sensor of this type are isolating the sensing and refer-

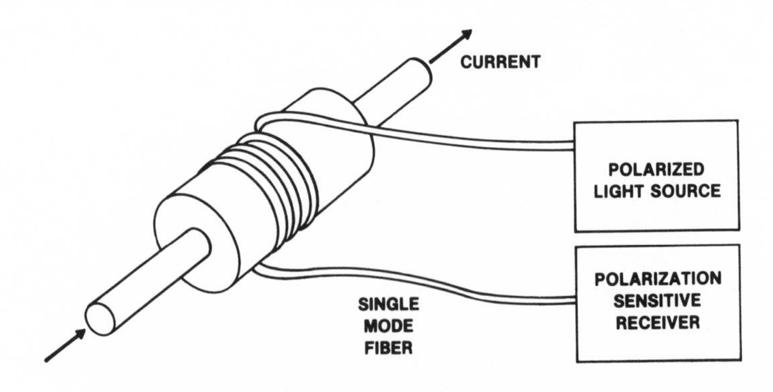

Figure 12-10 Faraday Rotation Current Sensor.

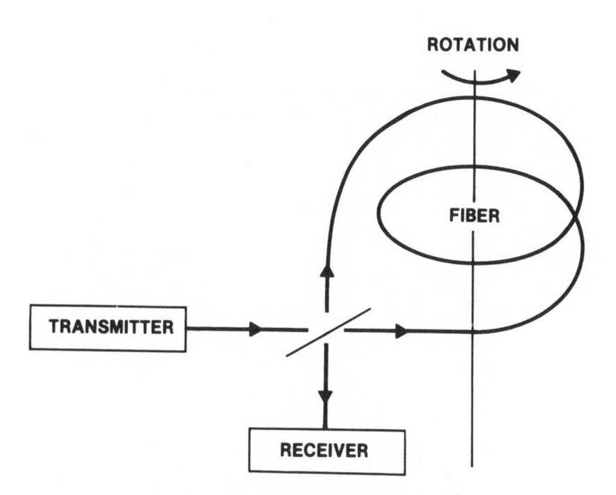

Figure 12-11 Rotation Sensor.

ence paths from extraneous environmental parameters (or making their effect on the phase delay the same for both paths), maintaining phase quadrature between the two paths in the presence of fluctuations of the optical source frequency, observing the changes in the detector output due to the sensed parameter in the presence of noise fluctuations from the optical source, and accommodating polarization rotations in the two paths, which modulate the interference terms.

Figure 12-10 shows a current sensor which operates on the Faraday rotation effect. Here a beam of polarized light, propagating in a fiber, travels parallel to the magnetic field produced by a current carrying conductor. This causes the polarization of the beam to rotate in proportion to the strength of the magnetic field. The polarization rotation is detected by a polarization sensitive optical receiver. It is interesting to note that this is actually a coherent sensor in disguise, where the two paths of the coherent sensor correspond to the two polarizations of propagation in the single mode fiber.

Figure 12-11 shows a coherent rotation sensor. Here a light signal is split into two parts which are launched in opposite directions through a coil of fiber. The signals emerge from their respective opposite ends of the fiber and are combined on a detector. Owing to a combination of ordinary and relativisitic effects, the phase delay around the fiber is not the same for the two directions of propagation, if the coil is rotating as shown. Thus the rate of rotation of the coil can be detected by the interference of the two combined optical signals. Rotation as small as 1/100 of the earth rate (1 rotation every 2400 hours) has been reliably measured by thus type of fiber optic gyroscope.

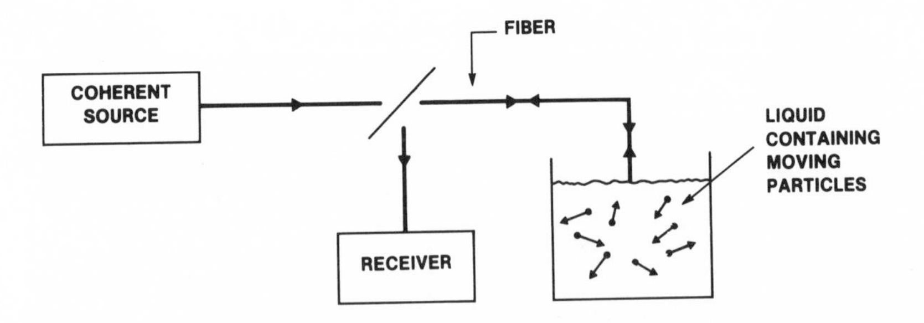

Figure 12-12 Particle Velocity Distribution Sensor.

Figure 12-12 shows a particle velocity distribution sensor. A light signal is launched into a fiber which illuminates a liquid containing moving particles. Some light is reflected directly from the tip of the fiber back toward a receiver as shown. Light from the myriad of moving particles is also reflected back into the fiber toward the receiver. The light reflected by the moving particles is Doppler shifted (in optical frequency). This light is mixed with the non-Doppler-shifted light at the detector (which responds to power) to produce components in the detector current which have a frequency distribution (electrical spectrum) corresponding to the distribution of particle velocities.

The above represent just a few interesting examples of the very large variety of sensing systems one can implement using optical fibers and electro-optic components. As with all sensing approaches, many practical problems which impact on the sensitivity, stability, and immunity to the effects of extraneous environmental parameters of these systems must be overcome by careful engineering methodology. However, the unique attributes of some of these fiber sensing approaches (including all-dielectric sensors without local power and extreme sensitivity, for example) make them very attractive.

13

System Measurements

There are several system measurements which are unique to fiber optic systems, as opposed to system measurements which are common to facilities implemented in other media (e.g., end-to-end bit-error-rate). We define a systems measurement as one which would be made during the installation, acceptance testing, or maintenance of a fiber optic system, as opposed to measurements which might be made in a factory as system components were being fabricated. In the sections below we shall discuss three of these systems measurements: end-to-end fiber loss, time domain reflectometry, and end-to-end bandwidth measurements.

13.1 End-to-End Fiber Loss Measurements

The typical approach for making an end-to-end loss measurement on an installed fiber is with a stabilized transmitter/receiver pair.

The transmitter can incorporate either an LED or a laser source (or several alternate source choices) emitting at the nominal wavelength of system operation (or wavelengths, for wavelength multiplexed systems or in anticipation of future wavelength multiplexing). The output power of the transmitter must be stable enough to obtain the desired accuracy of the end-to-end loss measurement. This is accomplished by a combination of temperature stabilization of the transmitter source and critical drive electronics, and/or feedback control of the light output using a local detector, as shown in Figure 13-1. The local detector can be attached to a passive coupler (tap) on the transmitter output pigtail, it can be placed on the back facet of a laser, or it can be positioned to pick up part of the light from an LED which is not coupled into the pigtail fiber. It must be remembered that the level of the light picked up by the local detector may not be perfectly correlated with the level of light launched, and that the degree of correlation depends upon the method used to obtain the locally

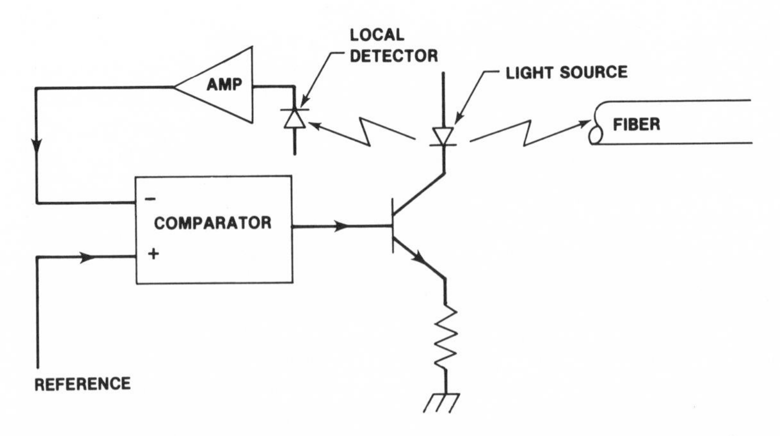

Figure 13-1 Test Set Transmitter with a Feedback Controlled Output Level.

detected and launched portions of the source output power. It is possible to obtain an output level stabilization of about 1% over an equipment ambient temperature range of several tens of degrees centigrade using these approaches. Output level stabilization is defined here as the variability of the total power launched into the transmitter pigtail, assuming that power launched into the pigtail cladding is removed by a suitable mode stripper. A stability of 1% corresponds to less than 0.05 dB of output level variability. However, one must be cautious about the much larger ambiguity associated with the coupling of the power from the calibrated/stabilized transmitter output pigtail into the fiber under test, both in terms of the total power coupled, and the way this power is distributed at launch amongst the guided modes of the fiber.

In some test transmitter designs, the output power is modulated by a low frequency sine or square wave oscillator so that the detected component of the modulated light at the fundamental frequency of the modulating waveform can be used as a reference. An example of such a transmitter is shown in Figure 13-2. This can reduce some problems with dc offsets in the transmitter and the complementary receiver, and also reduces problems associated with low frequency ($1/f$) noise in the receiver, and in the transmitter power stabilization control loop. However, to obtain the full benefits, one must carefully control the modulating frequency and waveform, and the gains of frequency selective amplifiers used in the receiver, and in the transmitter feedback control loop. One also requires a stable nonlinear element (rectifier or multiplier) in the receiver, following the bandpass filter, and in the transmitter control loop.

The receiver in the test set pair will typically incorporate a *p-i-n* detector and a transimpedance front-end amplifier. The *p-i-n* detector will have adequate sensitivity, since the loss measurement is made by averaging the received optical power level over a relatively long period of time, compared to the time scale of the modulation used when the communication system is in place. The *p-i-n* detector responsivity is very insensitive to temperature variations. The transimpedance amplifier is also relatively insensitive to temperature variations, provided the feedback resistor(s) used have a low temperature coefficient. The use of a low frequency modulation of the transmitter output and a corresponding bandpass filter in the receiver will help to control preamplifier current offset problems when the receiver is operating with very low power input levels.

The gain of the receiver can be precisely adjusted by using selectable resistors for the feedback element of the transimpedance amplifier, as shown in Figure 13-3. The capacitance associated with the wiring for the resistors and switches is not critical because of the low frequencies involved in the measurement process. The switches can be manually controlled or electronically controlled. With this approach, one can adjust the feedback resistance (by hand or via intelligent electronics in the receiver) until the received optical power produces a zero output from a comparator (a fixed "reading"). The relative level of the received optical power can then be deduced from the value of the adjusted feedback resistance. In addition to providing an excellent means for obtaining a precise and stable relative power measurement the use of a variable feedback resistance in the transimpedance amplifier results in a very large dynamic range for the receiver.

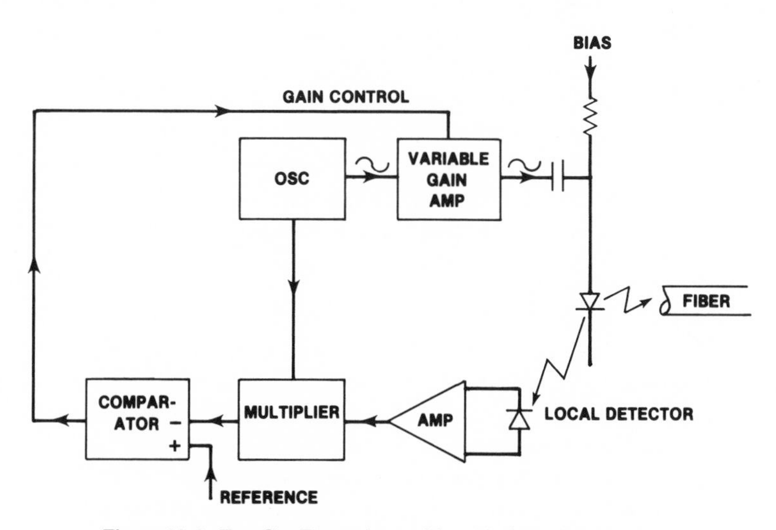

Figure 13-2 Test Set Transmitter with a Modulated Output Level.

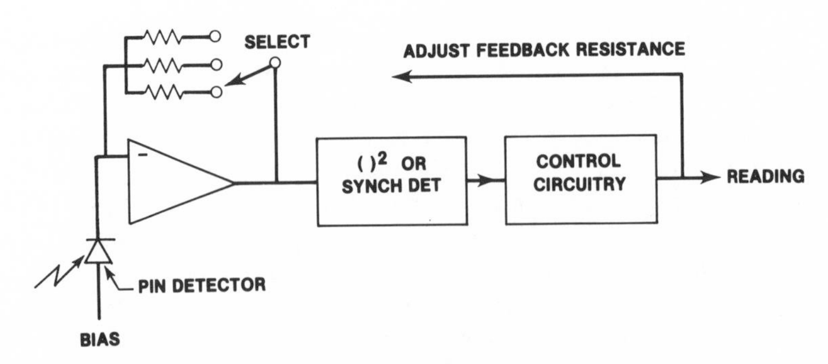

Figure 13-3 Receiver with Precision Adjustable Transimpedance.

The pigtail leading to the receiver is often selected to have a large core diameter and a large acceptance angle, in order to reduce the ambiguities associated with the coupling of power from the fiber under test to the pigtail. Using these approaches it is possible to construct a receiver with a relative accuracy of better than 1% and a stability of a fraction of 1% over a temperature range of several tens of degrees centigrade.

It is not necessary that the receiver have an absolute accuracy, if the power from the fiber output is going to be compared to the power emitted directly from the transmitter (or with some other reference). Relative accuracy refers to the accuracy of measurement of power received relative to power measured from some reference (e.g., the transmitter output).

Some test set receivers have absolute accuracy as well as relative accuracy so that they can be used as stand alone optical power meters. Modern test set receivers incorporate intelligent electronics to automatically adjust the feedback resistance, and provide a digital readout of the received optical power level or the fiber insertion loss.

It is not necessary that the output power level of the transmitter be accurately known (it should, however, be stable), because the receiver can referenced against the transmitter output before the fiber under test is measured (using a short fiber between the transmitter and the receiver).

13.2 Time Domain Reflectometers

A time domain reflectometer is an instrument which can be used to nondestructively measure the loss per unit length vs position along a fiber, or the insertion loss of a splice or connector. It can also be used to locate breaks and loss discontinuities in installed fiber cables.

The reflectometer is essentially an optical radar set as shown in Figure 13-4. A short duration pulse of light energy is launched into the fiber from an optical pulse transmitter, and reflections of this pulse vs time are observed via an optical receiver. Discrete reflections are obtained from breaks, non-perfectly-index-matched splices, and connectors. In addition, there is a continuum of reflected light vs time caused by Rayleigh scattering. As the launched pulse propagates down the fiber, a small fraction of its energy is continuously removed by the Rayleigh scattering mechanism in the glass (see Section 2.1.3 above). Of this Rayleigh scattered light, a small fraction is recaptured by the fiber in the reverse direction of propagation. The light received at any time T at the receiver (relative to the time the transmitter launches its pulse) corresponds to light which has been reflected from a position $cT/2n$ down the fiber, where c/n is the speed of light in the fiber core, and the factor of 2 takes into account the round trip. In traveling to position L along the fiber, the transmitted pulse is attenuated by the fiber loss. Similarly, the returning scattered or reflected light is also attenuated. Thus the observed backscattered light vs time has a decaying shape which is approximately a replica of the cumulative round-trip loss vs distance along the fiber. Figure 13-5 shows a typical trace. By observing the rate of decay of the scattering portion of this

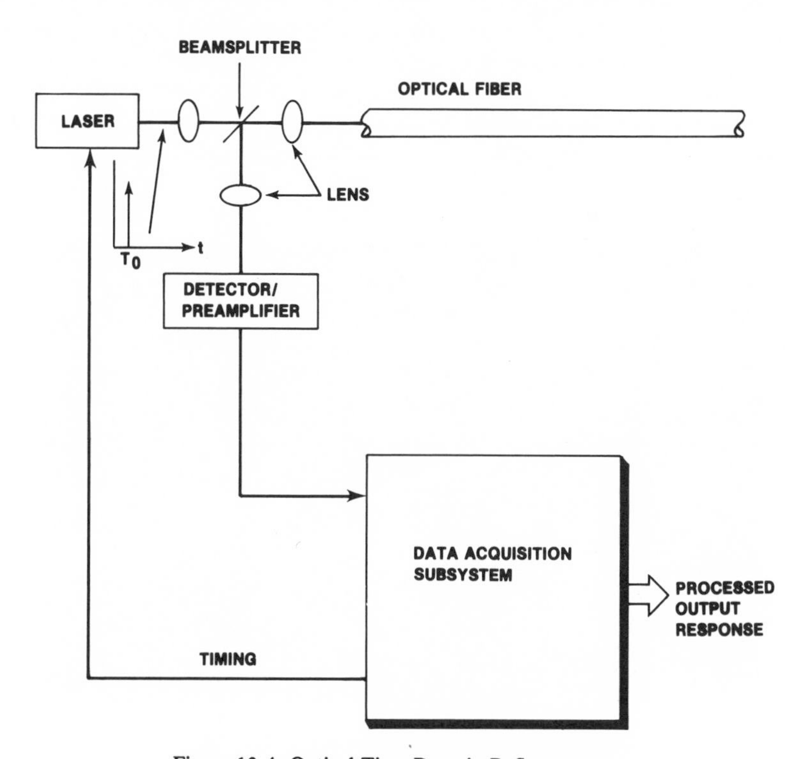

Figure 13-4 Optical Time Domain Reflectometer.

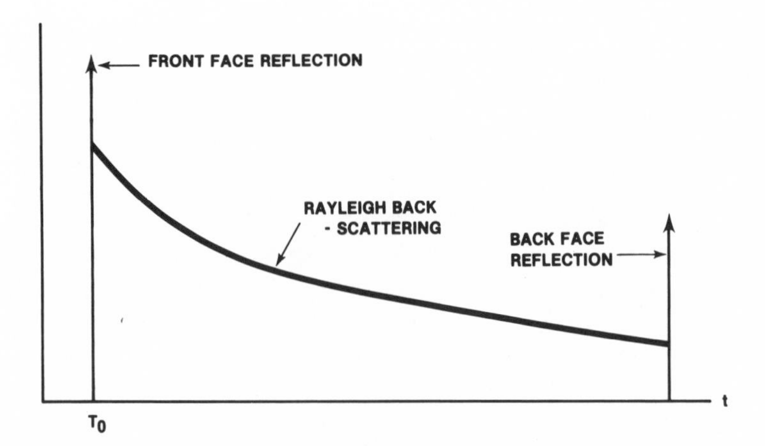

Figure 13-5 Rayleigh Scattering OTDR Trace.

trace, one can estimate the loss of the fiber vs position, and the round trip losses of connectors and splices. Figure 13-6 shows an actual trace taken from a fiber approximately 1500 m long, with a high loss region at the far end. One can see the rapid drop off of the Rayleigh scattering just before the large reflection from the unterminated far end.

The Rayleigh scattered light is very low in amplitude, even for a very powerful transmitter. This is even more of a problem with single mode

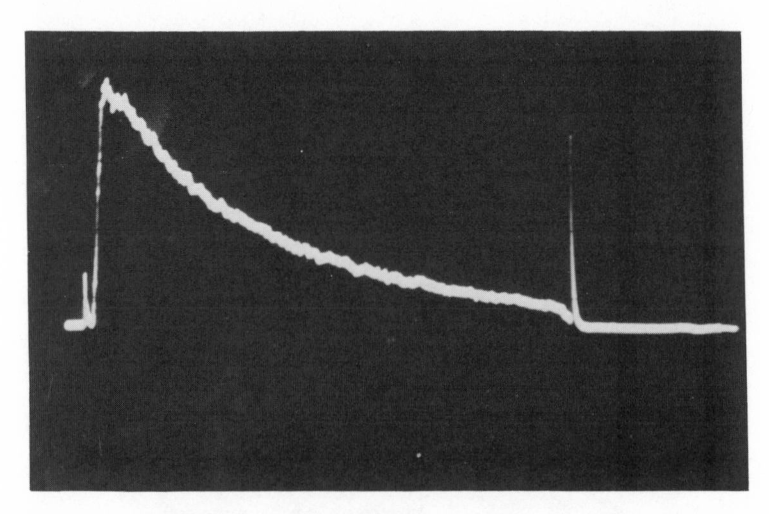

Figure 13-6 Actual OTDR Trace from a 1500-m-Long Fiber. Reprinted with Permission of Bell Communications Research and AT&T Bell Laboratories.

fibers, since they capture much less of the Rayleigh scattered light in the backward propagating direction compared to multimode fibers. For this reason, the distance one can "look down" a fiber is typically limited to few few tens of decibels of one-way loss. The specific limit depends upon the sophistication of the receiver and the energy launched in each transmitted pulse.

Optical time domain reflectometers are considered to be an indispensible tool for characterizing cables, fibers, splices, and connectors during installation and as part of ongoing maintenance.

13.3 Measurements of Fiber Bandwidth

The bandwidth of an optical fiber can be limited by pulse spreading caused by modal delay spread or material dispersion, or by mode partition noise.

For multimode fibers, pulse spreading is typically measured by launching a temporally narrow pulse of light into the fiber and observing the shape of the output pulse. If the transmitter used in this test has a narrow optical spectrum compared to the transmitter which will be used to carry information in the working system, then one may have to analytically adjust the material dispersion component of the spreading observed in the test.

A typical difficulty with bandwidth measurements in multimode fibers is obtaining a repeatable result. This problem arises for two reasons. First, the pulse spreading in a multimode fiber is very sensitive to the distribution of energy among the fiber modes at launch. Second, when the pulse spreading is small, it is difficult to obtain a fast enough transmitter and receiver to make the back-to-back transmitter—receiver response narrow in time compared to the fiber pulse spreading effect. If one tries to observe a small change in the width of the receiver output pulse, with and without the fiber-under-test between the transmitter and the receiver, then nonlinearities in the receiver and/or noise in the transmitter and the receiver (including pickup) make this very difficult to do accurately.

An alternative to using pulses is to modulate the transmitter with a sinusoidal signal of variable frequency. The frequency response of the back-to-back transmitter/receiver pair is compared to the frequency response of the pair with the fiber inserted between them. This measurement suffers from the same shortcomings described above for pulse measurements, but may be easier to implement for peak power limited optical sources.

Compounding the above problems is the difficulty in extrapolating measurements made of the frequency responses or bandwidths of individual fibers to the measured frequency responses or bandwidths of concatenations of these same fibers. This occurs not only because of the individual fiber measurement difficulties, but also because of the complex nature of the relationship between the responses of individual and concatenated fibers having real-world index profile variabilities and unpredictable mode coupling at splices.

As a result of all of this, bandwidth measurements made on installed multimode fibers generally provide only an approximate insight into the information-carrying capabilities of those fibers.

In single mode fibers, the dispersion of the fiber is the bandwidth limiting parameter of interest. Dispersion is often measured by comparing the propagation delay down the fiber for two sources at different wavelengths near the nominal wavelength of interest. This can be accomplished by launching two short duration pulses into the fiber (at two wavelengths) and observing how the temporal separation between them changes, between the fiber input and the fiber output. As an alternative, two laser sources at different wavelengths can be modulated with sinousoids of variable frequency, and the associated relative delays (of the two wavelengths) down the fiber can be inferred by observing the received sinousoidally modulated waveforms with phase sensitive detectors (the slope of the phase vs frequency characteristic is proportional to the delay through the link).

Unlike the pulse spreading measurement in multimode fibers, the dispersion measurement in single mode fibers is a reproducible and important measure of the fiber's information-carrying capability with various types of optical sources (various source spectral characteristics).

14

Emerging Technology and Applications

There are a number of emerging technologies, and applications of those technologies, which may be important in the future. These include integrated optoelectronics, heterodyning systems, and optical (photonic) switching.

14.1 Integrated Optoelectronics

In Chapter 5 we described devices such as modulators, couplers, wavelength multiplexers, and switches, which could be fabricated in the surface of a passive or an electro-optically active material using diffusion, ion exchange, or other techniques. It is possible to fabricate these devices in materials which can also generate, amplify, and detect light (e.g., indium phosphide). Furthermore it is also possible to fabricate electronic components such as field effect transistors in these same materials. If a practical technology for fabricating all of these optical, electro-optical, and electronic components on a monolithic chip could be developed, then one could, in principle, manufacture very low cost, highly functional, transmitters and receivers which could operate at very high data rates.

Figure 14-1 shows an example of an integrated optics chip which contains only optical switches and which acts as a multiplexer/modulator. Starting at the left, a laser generates an unmodulated light signal which is coupled into waveguide 1, as shown. Waveguides 1 and 2 pass under a pair of electrodes which can be used to modulate the coupling between these two contiguous waveguides. Using this modulator, one can periodically switch the laser output power between the two waveguides as shown, using a square wave signal to drive the electrodes. In this example we assume that the square wave repetition rate is 1 Gb/s. Waveguides 1 and 2 carry their square wave modulated (out of phase) light signals to another pair of

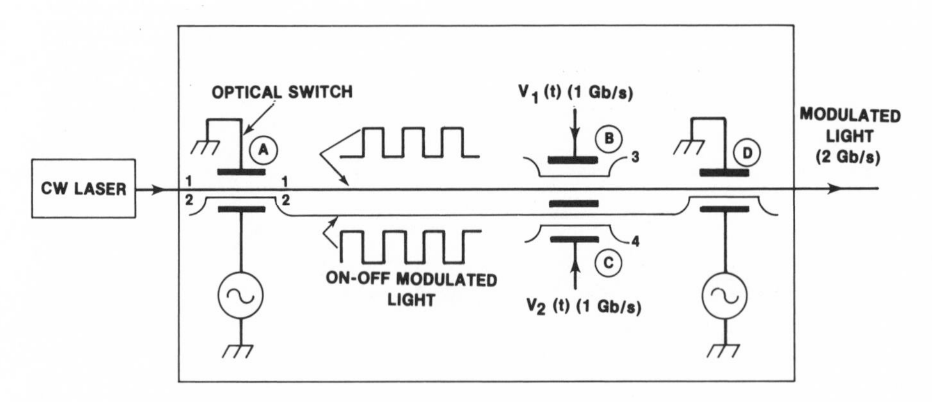

Figure 14-1 Integrated Optics Multiplexer/Modulator.

modulators. Modulator B can vary the coupling between waveguides 1 and 3. Modulator C can vary the coupling between waveguides 2 and 4. Using separate digital information signals, each at 1 Gb/s rate, we modulate the pulse streams emerging from these modulators in waveguides 1 and 2-respectively. Each 1 Gb/s digital modulating waveform determines which of the pulses are passed and which are diverted in the respective modulators. (The input to each of these modulators is a repetitive 1 Gb/s stream of half duty cycle pulses.) These two individually modulated optical pulse streams are recombined (interleaved) in the final modulator D. Thus we create a composite optical bit stream at 2 Gb/s from two electrical bit streams, each at 1 Gb/s. Using similar techniques, we could separate a high rate optical bit stream into several parallel bit streams, each at a correspondingly lower rate.

By fabricating multiple sources at different wavelengths on an integrated optoelectronic chip, along with modulator electronics and a wavelength multiplexer we could implement a very low cost transmitter with a very high composite output data rate (e.g., 1 Gb/s on each of four wavelengths). Similarly, we could fabricate a receiver chip containing a wavelength demultiplexer and several optical receivers.

These are just a few examples of what might be possible with emerging integrated optoelectronic technology.

14.2 Heterodyning (Coherent) Technology and Applications

In the application examples in Chapters 7—12 above (except coherent sensor systems) we modulated the output power of an optical source, and we detected the modulated optical power with a *p-i-n* or avalanche photodiode. It is also possible to modulate the phase of an optical signal and to detect the phase of a received optical signal using coherent detection techniques. Even if we modulate only the amplitude of an optical signal, coherent detection techniques can be used to implement a receiver with a very narrow optical bandwidth and an improved sensitivity.

The principle of coherent detection is shown in Figure 14-2. An incoming optical signal is added to a local oscillator, having a well defined frequency or phase relationship with the incoming signal. The local oscillator must also have the same spatial field pattern and polarization as the incoming signal. The sum of these two signals falls upon a optical power detector, which produces an output current proportional to the total power incident upon it. If the incoming light field is given by

$$E_{in} = \sqrt{2}\ \mathrm{Re}[m(t)\ \Psi(x,y)\ \exp(ift)] \tag{14-1}$$

where f is the incoming optical frequency (rad/sec), and $m(t)$ is a message containing modulation, $\Psi(x,y)$ is the spatial field pattern, and $\mathrm{Re}(x)$ signifies "real part of x," and if the local oscillator field is given by

$$E_{\mathrm{l.o.}} = \sqrt{2}\ \mathrm{Re}[A\ \Psi(x,y)\ \exp(\mathrm{i}f't + \theta)] \tag{14-2}$$

where f' is the local oscillator optical frequency, and θ is a slowly varying local oscillator phase term, then the detector current response is given by

$$i_d(t) = \eta \frac{e}{hf} I_o \{|m(t)|^2 + |A|^2 + 2A\, m(t)\ \cos[(f-f')t-\theta]\} \tag{14-3}$$

where

$$I_0 = \int_{\text{detector area}} |\Psi(x,y)|^2\, dx\, dy$$

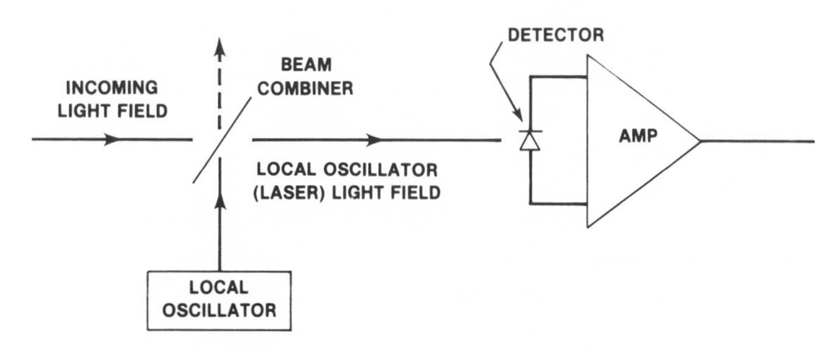

Figure 14-2 Coherent Detection Receiver.

Note that the cross term is proportional to the amplitude of the incoming field $m(t)$.

One can show that the noise at the output of a coherent detection receiver will be dominated by shot noise from the local oscillator (amplifier thermal noise is negligible). Further, provided that the local oscillator and the incoming light signal maintain a reasonably controlled frequency difference, the local oscillator and incoming light signal spatial field patterns are reasonably well matched, and the local oscillator and the incoming light signal have reasonably similar polarizations, one can obtain "quantum limited" receiver sensitivity performance. Recall that for digital on—off modulation, and a receiver incorporating a *p-i-n* detector and employing direct detection, the receiver sensitivity is about 27 dB worse than the "quantum limit" (12,000 photons per pulse vs the direct-detection quantum limit of 21 photons per pulse, for an error rate of 10^{-9}). In a direct detection receiver employing a silicon or InGaAsP (at long wavelengths) avalanche detector, the receiver sensitivity is within about 10—20 dB of the direct-detection quantum limit (200—2000 photons per pulse) depending upon the details of the detector and the receiver. The quantum limited performance of an ideal heterodyne receiver for on—off digital modulation is 72 photons per pulse. Thus the improvement in receiver sensitivity performance with *ideal* heterodyning, compared to *practical* direct detection with an avalanche photodiode, may be 5—15dB.

To implement practical heterodyning requires specially stabilized light sources and single mode technology. Tracking loops are needed to maintain a well controlled frequency offset between the local oscillator and the incoming light signal, as well as to compensate for random polarization changes.

The implementation of a transmitter with a well controlled optical frequency may imply the necessity of using an external modulator (since direct modulation may cause frequency modulation of the transmitter source, and since the stabilization techniques may preclude direct modulation at the desired speeds). It is possible that the insertion loss of the external modulator, the insertion loss of the beam combiner at the receiver, and other deviations from ideal implementation may offset some or all of the improved sensitivity (vs direct detection) associated with a heterodyne (coherent) receiving subsystem.

Heterodyning is complex to implement with discrete optical and optoelectronic components but may be practical to implement with emerging integrated optics technology.

Figure 14-3 shows an example of an application for heterodyning where a receiver selects one of several closely spaced wavelengths from a multiwavelength fiber loop.

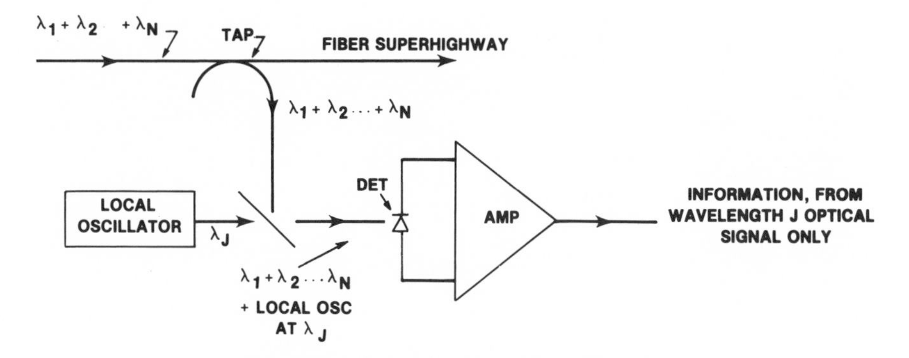

Figure 14-3 Example of Heterodyning Used in a System.

Thus one possible application for coherent optical technology is in approaching the ultimate bandwidth potential of single mode fibers with large numbers of closely spaced, separately modulated, optical carriers. However, there is some concern that the total power associated with all of these optical carriers may initiate undesirable non-linear effects in single mode fibers. This subject is under investigation both analytically and experimentally.

14.3 Photonic Switching

With fiber optic and optoelectronic technology having such a big impact on reducing costs and adding new capabilities in telecommunication transmission facilities, one might reasonably ask what impact this same technology might have in telecommunication switching applications. If signals arrive at a switching facility in optical form, why not leave them in that form and use an optical switching matrix?

The answer to this question is somewhat dependent upon the application, but in general it is difficult to implement large switching matrices, which imitate the architecture of today's electronic switching systems, with present and emerging optoelectronic technology. A large array of optical switches, like those discussed in Section 5.2 above, would introduce significant insertion loss and crosstalk between optical signals propagating through the matrix. Thus a large matrix would have to include not only optical crosspoints, but some means for optical regeneration. If we include optical regeneration (including retiming) in the switching matrix we may have to convert the optical signals to electrical form to accomplish the regeneration process. (However, all optical regeneration might be possible.) If we do this, one might ask: once you have the signal in electrical form, why not do the switching electronically as well?

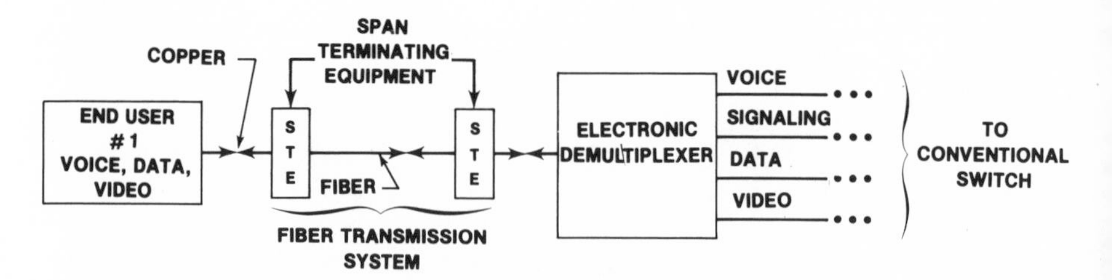

Figure 14-4 Multiservice Switching Example.

In telecommunications applications, the signal arriving at the switch typically will contain a multiplex of information, including call setup information, which must be routed to different places. To separate these multiplexed information signals one could use an optical-to-electronic conversion process as shown in Figure 14-4. Again one may ask: with the signal converted to electrical form for demultiplexing, why not switch it electronically?

The answer to this question is generally that we should indeed implement the switching function electronically with today's technology and today's switching system architectures. However, in the future, as new services and network architectures lead to higher and higher rate optical signal streams, and as integrated optics technology provides a richer class of optical signal processing functions, we may find that all optical switching (photonic switching) will become a practical reality. Note that as in the case of LANs (Chapter 9) the best utilization of optical components in

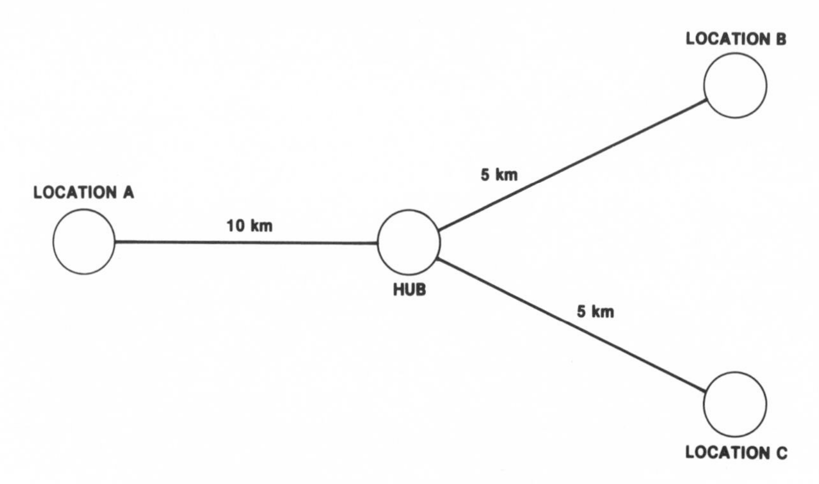

Figure 14-5 Mechanical Optical Switch Application.

switching may require innovations in the subsystem/network architecture as well as in the components themselves.

For the implementation of relatively small switching matrices (e.g., eight inputs and eight outputs) one can use less exotic technologies, including mechanical optical switches as described in Section 5.1 above. In this type of application an all optical switch offers the advantage of being literally transparent to the wavelength of operation and the modulation rate or format being used.

Figure 14-5 shows an example of a small network where a small optical switch placed in the hub location allows any pair of locations to communicate. If only one pair of locations needs connectivity at any one time, then such an approach can save on fiber costs in exchange for the (hopefully lower) cost of a remotely controllable optical switch.

In summary, small switching matrices, not requiring rapid reconfiguration, could be configured with today's mechanical optical switching technologies. Small switching matrices requiring rapid reconfiguration could be configured with emerging integrated optics electrically controlled crosspoints. The implementation of large all-optical switching matrices requires the invention of all-optical regeneration and retiming functions, and their incorporation into large integrated optoelectronic circuits.

References

This manuscript summarizes the contributions of many researchers to whom the scientific community and this author in particular are greatly indebted. The small cross section of references below was selected as most appropriate to guide the reader toward more in-depth study of the subjects presented in this book. An excellent overview of the historical evolution of the art with an extensive list of references can be found in Reference 13 below.

Historical References

[1] Kao, K. C., and Hockham, G. A., Dielectric surface waveguides for optical frequencies, *Proc. IEE* **133**, 1151-1158 (July, 1966).

[2] Kapron, F. P., Keck, D. B., and Maurer, R. D., Radiation losses in glass optical waveguides, *Appl. Phys. Lett.* **17**, 423-425 (November, 1970).

[3] Miya, T., Terunuma, Y., Hosaka, T., and Miyashita, T., Ultimate low-loss single mode fiber at 1.55 μm, *Electr. Lett.* **15**, 106-108 (February 1979).

[4] Gloge, D., Weakly guiding fibers, *Appl. Opt.* **10**, 2252-2258 (October 1971).

[5] Gloge, D., and Marcatili, E. A. J., Multimode theory of graded core fibers, *Bell Syst. Tech. J.* **52**, 1563-1578 (November 1973).

[6] Burrus, C. A., and Dawson, R. W., Small-area high-current-density GaAs electroluminescent diodes and a method of operation for improved degradation characteristics, *Appl. Phy. Lett* **17**, 97-99 (August 1970).

[7] Hayashi, I., Panish, M. B., Foy, P. W., and Sumski, S., Junction lasers which operate continuously at room temperature, *Appl. Phy. Lett.* **17**, 109-111 (August 1970).

[8] Personick, S. D., Receiver design for digital fiber optic communication systems, *Bell Syst. Tech. J.* **52** (6), 843-886 (July-August 1973).

[9] Goell, J. E., A 274 Mb/s repeater experiment employing a GaAs laser, *Proc. IEEE* **61**, 1504-1505 (October 1973).

[10] Sell, D., and Maione, T., Experimental fiber optic transmission system for interoffice trunks, *IEEE Tran. Comm.* **COM-25** (5), 517-522 (May 1977).

[11] Atlanta Experiment, *Bell Syst. Tech. J.* **57** (6), Part 1 (July-August 1978).

[12] Miller, S. E., Marcatili, E. A. J., and Li, T., Research toward optical fiber transmission systems, *Proc. IEEE* **61**, 1703-1751 (December 1973).

[13] Li, T., Advances in optical fiber communication — an historical perspective, *IEEE J. on Selected Areas in Comm.* **SAC-1** (3), 356-372 (April 1983).

Books

[14] Personick, S. D., *Optical Fiber Transmission Systems*, Plenum Press, New York (1981).

[15] Miller, S. E., and Chynoweth, A. G., *Optical Fiber Telecommunications*, Academic Press, New York (1979).

[16] Kressel, H., *Semiconductor Devices for Optical Communications*, Springer-Verlag, Berlin, Germany (1980).

[17] Barnoski, M. K., *Fundamentals of Optical Fiber Communications, 2nd Edition*, Academic Press, New York (1981).

[18] Technical Staff of CSELT, *Optical Fiber Communications*, CSELT, Torino, Italy (1980).

[19] Midwinter, J. E., *Optical Fibers for Transmission*, Wiley, New York (1979).

[20] Kao, K. C., *Optical Fiber Systems — Technology, Design, and Applications*, McGraw Hill, New York (1982).

[21] Keiser, G., *Optical Fiber Communication*, McGraw Hill, New York (1983).

[22] Culshaw, B., *Optical Fiber Sensing and Signal Processing*, Peter Peregrinus Ltd., London, United Kingdom (1984).

Special Journal Issues

[23] *IEEE Trans. Comm.*, Special Issue on Fiber Optics, **COM-26** (7), (July 1978).

[24] *IEEE J. of Selected Areas in Comm.* Special Issue on Fiber Optic Systems **SAC-1** (3), (April 1983).

[25] *IEEE J. of Lightwave Technology*, Special Issue on Undersea Cable Fiber Optic Systems **LT-2** (6), (December 1984).

[26] *IEEE J. of Lightwave Technology*, Special Issue on Local Area Networks, **LT-3** (3), (June 1985).

Current Journals

[27] *IEEE Journal of Lightwave Technology*

[28] *IEEE Journal of Quantum Electronics*

[29] *IEEE Transactions on Communications*

[30] *IEEE Journal of Selected Areas in Communications*

[31] *Electronics Letters* (IEE, United Kingdom)

Conference Proceedings

[32] *Proceedings of OFC*, 1975, 1977, 1979, 1981–, Optical Society of America, Washington, D. C.

Recent Technical Publications Referenced

[33] Shi, Y. C., Klafter, L., and Harstead, E. E., Dual-channel bidirectional optical rotary joint, *Proc. OFC85* (poster paper), San Diego, February 11-13, 1985.

[34] Bates, R. J. S., et al., 1.3 μm/1.5 μm Bidirectional WDM optical-fiber transmission system experiment at 144 Mb/s, *Electron Lett.*, **19** (13), 458-459 (June 1983).

Index